STEM Inquiry and Its Practice in K-12 Classrooms

Through examining the theoretical ideas of disciplinarity and disciplinary practices, the book presents instructional aspects for teachers to explore when engaged with integrated STEM inquiry.

Are you interested to understand the difference between science inquiry and STEM inquiry? Do you want to introduce integrated STEM problem solving to your students but need help with the key features of STEM inquiry? This book presents in-depth discussions related to the features and affordances of integrated STEM inquiry. Written for K-12 teachers and teacher educators, this book conceptualises STEM inquiry and integrated STEM and their enactment, using three practical STEM instructional frameworks: problem-centric, solution/design-centric, and user-centric STEM. The three STEM instructional frameworks serve as a key anchor for teachers to interpret and apply when planning various STEM lessons in meaningful, practical, and coherent ways.

Whether you are an aspiring K-12 STEM teacher or an in-service teacher teaching K-12 students, the ideas of integrated STEM inquiry presented in this book challenge educators to think about the principles of integrated STEM inquiry and how they can be incorporated into classroom practice and lessons.

Aik-Ling Tan is Deputy Head (Teaching & Curriculum Matters), Natural Sciences and Science Education. She is also an associate professor of science education at the National Institute of Education, Nanyang Technological University, Singapore. She teaches biology education methods courses and courses related to integrated STEM curriculum. Prior to teaching at the National Institute of Education in 2007, Aik-Ling taught Biology and Lower Secondary General Science at River Valley High School, Singapore for ten years.

Tang Wee Teo is an associate professor in the Natural Sciences and Science Education at the National Institute of Education, Nanyang Technological University, Singapore. She is also the Co-Head of the Multi-centric Education, Research and Industry STEM Centre (meriSTEM@NIE).

Jina Chang is a visiting scholar at the National Institute of Education, Nanyang Technological University, Singapore. Her research interests include science inquiry from a sociocultural perspective. She has recently broadened her interests to encompass STEM education and multimodal approach in science inquiry.

Ban Heng Choy is an assistant professor in Mathematics Education at the National Institute of Education, Nanyang Technological University, Singapore. As a recipient of the NIE Overseas Graduate Scholarship, Dr Choy received his PhD from the University of Auckland, New Zealand, in 2015. He is currently one of the co-Heads for meriSTEM@NIE, a Multi-centric Education, Research, and Industry STEM centre in NIE.

Routledge Research in STEM Education

The *Routledge Research in STEM Education* series is home to cutting-edge, upper-level scholarly studies and edited collections covering STEM Education.

Considering science, technology, engineering, and mathematics, texts address a broad range of topics including pedagogy, curriculum, policy, teacher education, and the promotion of diversity within STEM programmes.

Titles offer dynamic interventions into established subjects and innovative studies on emerging topics.

Teaching Assistants, Inclusion and Special Educational Needs:
International Perspectives on the Role of Paraprofessionals in Schools
Edited by Rob Webster and Anke A. de Boer

Invention Pedagogy – The Finnish Approach to Maker Education
Edited by Tiina Korhonen, Kaiju Kangas, and Laura Salo

Visualisation and Epistemological Access to Mathematics Education in Southern Africa
Edited by Marc Schäfer

Recruiting Black Biology Majors into STEM Education Careers
Journeys to Success
Salika A. Lawrence, Tabora A. Johnson, and Chiyedza Small

Post-Secondary Chemistry Education in Developing Countries
Advancing Diversity in Pedagogy and Practice
Dawn I. Fox, Medeba Uzzi and Jacqueline Murray

STEM Inquiry and Its Practice in K-12 Classrooms
Activities, Characteristics, and Enactment
Aik-Ling Tan, Tang Wee Teo, Jina Chang, and Ban Heng Choy

For more information about this series, please visit: www.routledge.com/Routledge-Research-in-STEM-Education/book-series/RRSTEM

STEM Inquiry and Its Practice in K-12 Classrooms
Activities, Characteristics, and Enactment

Aik-Ling Tan, Tang Wee Teo, Jina Chang, and Ban Heng Choy

LONDON AND NEW YORK

First published 2024
by Routledge
4 Park Square, Milton Park, Abingdon, Oxon OX14 4RN

and by Routledge
605 Third Avenue, New York, NY 10158

Routledge is an imprint of the Taylor & Francis Group, an informa business

© 2024 Aik-Ling Tan, Tang Wee Teo, Jina Chang, and Ban Heng Choy

The right of Aik-Ling Tan, Tang Wee Teo, Jina Chang, and Ban Heng Choy to be identified as authors of this work has been asserted in accordance with sections 77 and 78 of the Copyright, Designs and Patents Act 1988.

All rights reserved. No part of this book may be reprinted or reproduced or utilised in any form or by any electronic, mechanical, or other means, now known or hereafter invented, including photocopying and recording, or in any information storage or retrieval system, without permission in writing from the publishers.

Trademark notice: Product or corporate names may be trademarks or registered trademarks, and are used only for identification and explanation without intent to infringe.

British Library Cataloguing-in-Publication Data
A catalogue record for this book is available from the British Library

ISBN: 978-1-032-72757-8 (hbk)
ISBN: 978-1-032-72769-1 (pbk)
ISBN: 978-1-003-42250-1 (ebk)

DOI: 10.4324/9781003422501

Typeset in Times New Roman
by SPi Technologies India Pvt Ltd (Straive)

Contents

List of Figures   *ix*
List of Tables   *xi*
Preface   *xii*
Acknowledgements   *xvi*

1 **Making sense of STEM inquiry: Reasons, inquiry, and problem solving**   1
  1.1 *Exploring teachers' conception of STEM learning* 2
  1.2 *Understanding STEM disciplinary features* 4
  1.3 *Understanding integration models* 15
  1.4 *Role of problems and problem solving in STEM* 17

2 **Planning for integrated STEM**   28
  2.1 *How science and engineering are interwoven in NGSS* 28
  2.2 *The STEM Quartet instructional framework of integrated STEM learning* 31
  2.3 *Mapping STEM practices to disciplinary expertise* 37
  2.4 *Mapping specific STEM practices* 38

3 **Disciplinarity in integrated STEM inquiry**   42
  3.1 *Discovering disciplinarity* 42
  3.2 *Variations of integrated STEM inquiry* 54

4 **Assessing integrated STEM inquiry**   57
  4.1 *An assessment model for integrated STEM inquiry* 57
  4.2 *Assessing* what: *Deciding on the categories of an assessment* 59
  4.3 *Assessing* how: *Developing a suite of different assessment methods* 63
  4.4 *A formative and iterative process of assessing integrated STEM inquiry* 67

5 Enactment of integrated STEM inquiry in classrooms 72
5.1 Flexibility in integrated STEM inquiry enactment 72
5.2 Conditions for integrated STEM inquiry enactment 73
5.3 Hybrid applications of implementing integrated STEM inquiry 75

6 Working as a multidisciplinary team 79
6.1 Vignette 1—Working together as a teaching team 82
6.2 Relating vignette to conditions for successful integrated STEM inquiry 88

7 Learning from nature's design 92
7.1 Vignette 2—Biomimicry 95
7.2 Relating vignette to conditions for successful integrated STEM inquiry 99

8 Growing humanistic values through integrated STEM inquiry 102
8.1 Vignette 3—Humanistic STEM 104
8.2 Relating vignette to conditions for successful integrated STEM inquiry 115

9 Working with evidence in integrated STEM inquiry 119
9.1 Vignette 4—Harnessing data 121
9.2 Relating the vignette to conditions of successful integrated STEM inquiry 128

10 Using expert voices to increase task authenticity 131
10.1 Vignette 5—Improving lives 133
10.2 Relating the vignette to the characteristics for successful integrated STEM inquiry 134

11 Conclusions 137

Index 140

Figures

1.1	Teachers' conceptual models of STEM education (a) STEM as an acronym (b) real-world problem solving as context (c) science as context (d) science, technology, engineering, and mathematics as separate disciplines (e) integrated disciplines (f) engineering design process as context (g) science and engineering design process as context (h) engineering as context (Ring et al., 2017)	3
1.2	Guided discovery to investigative approach (Yeo & Choy, 2023)	11
1.3	Conceptualisations of STEM inquiry based on general inquiry, science, and mathematical inquiry	12
1.4	Students bring in and integrate STEM disciplines with the problem situation (Bybee, 2019)	17
1.5	Conceptual framework for STEM learning (Kelley & Knowles, 2016)	18
1.6	S-T-E-M Quartet instructional framework (Tan et al., 2019)	19
1.7	Connections across disciplines in an integrated STEM activity (Tan et al., 2023a)	23
2.1	The problem-centric STEM learning process (Tan et al., 2019)	33
2.2	The solution-centric STEM learning process (Teo et al., 2021)	34
2.3	The user-centric STEM learning process (Teo et al., 2021)	35
4.1	A competence-based assessment integrating three key areas	60
4.2	Assessment methods used in STEM education to evaluate student learning domains (Cheung & Yeh, 2022)	64
4.3	A formative and iterative process of STEM practice assessment	68
4.4	An example of teacher-subjective assessment of problem-solving skills	69
4.5	An example of peer-reported assessment of collaboration in a round-robin activity	70
6.1	Representing wastewater treatment lesson using the STEM Quartet instructional framework	80

x *Figures*

6.2	An example of the source code	83
6.3	Understanding flowrates	84
6.4	One of the three tableful of tools	85
6.5	Mounting board that was shared	86
6.6	Troubleshooting with students	87
6.7	Checking for leaks	88
7.1	(a) Representing biomimicry design of litter pick	93
7.1	(b) Representing biomimicry design of foldable raft	94
7.2	An example of a prototype to pick litter	97
7.3	A crab claw prototype	97
7.4	Testing prototype	98
8.1	Mapping learning outcomes of user-centric STEM	103
8.2	Using videos to raise students' awareness	104
8.3	Group discussion using graphic organiser	105
8.4	Whole-class discussion	106
8.5	Students sharing their ideas documented on their thinking board	106
8.6	Learning tasks for graph theory	107
8.7	Challenge problem for the STEM unit	108
8.8	Students modelling the problem using graph theory	109
8.9	Students working on the 'water transfer speed challenge'	110
8.10	Students working on the experiments	111
8.11	Mr Teerute relating the Bernoulli principle to the village's solution	112
8.12	The scaled model building task	113
8.13	Students building scaled model of the mountain	114
8.14	Students discuss the design using the scaled model	115
9.1	An integrated STEM lesson on designing a pigeon repellent using the STEM Quartet instructional framework	120
9.2	Students presenting their mind map and ideas to their peers	122
9.3	Students at a learning station on sound reception	123
9.4	Flipchart showing the pigeon-repellent design proposed by a group of students	124
9.5	Students getting feedback on their pigeon-repellent design from their peers	125
9.6	Material and tools for constructing pigeon-repellent prototypes	126
9.7	Students using Geometer's Sketchpad to design a model of their prototype	127
9.8	Students presenting their Geometer's Sketchpad model	128
10.1	An integrated STEM lesson on improving the lives of rubber tappers using the STEM Quartet instructional framework	132

Tables

1.1	Goals and disciplinary practices in the STEM disciplines (Bybee, 2019)	5
1.2	Different levels of integration	16
2.1	Differences in the practices of science and engineering (Cunningham & Carlsen, 2014)	30
2.2	Comparison of the problem-, solution-, and user-centric variants (Teo et al., 2021)	36
2.3	Variations of science-, technology-, engineering-, and mathematics-led practices mapped to different centricities	39
2.4	An example of mapping STEM practices to a solution-centric lesson	40
3.1	A framework for integration of STEM pedagogical practices for lesson enactment for teachers (Ong et al., 2023)	44
3.2	Mapping conditions of successful STEM inquiry to centricities	53
4.1	Three categories of assessment in STEM learning	60
4.2	Examples of integrated forms of competencies within the context of STEM problems	61
4.3	Using a rubric to determine assessment categories in integrated STEM inquiry	62
4.4	An example of rubric for individual student self-reported assessment	65
4.5	An example of rubric for group peer-reported assessment	66
4.6	An example of rubric for teacher-subjective group assessment	66
4.7	Categorising and developing a suite of different STEM assessment methods	67
5.1	A hybrid application of the three variants of the STEM Quartet instructional framework (4Cs × Problem-Solution-User)	77
6.1	Understanding the variation in practice for vignette 1	91
7.1	Understanding the variation in practice for vignette 2	101
8.1	Understanding the variation in practice for vignette 3	117
9.1	Understanding the variation in practice for vignette 4	130
10.1	Understanding the variation in practice for vignette 5	136

Preface

The release of the Next Generation Science Standards (NGSS) in April 2013 hastened efforts in integrating engineering ideas, technological advances, and mathematical thinking into science teaching and learning. In the history of mankind, scientific and mathematical knowledge has traditionally been acknowledged as important cultural and intellectual achievements (PISA, 2023). Learning about the knowledge and skills in science and mathematics has long been touted as important in school curricula across the world. Generations of students have memorised, learnt, and recapitulated the knowledge of science and mathematics to pass milestone entrance or placement examinations to earn a place in college (for a course of study that may not be science or mathematics). Despite the well-established knowledge base and epistemic practices of science and mathematics, disciplinary reforms have led to calls for science and mathematics learning to be made more relevant to students' lives and the society at large. These calls have resulted in attempts to re-design learning experiences that are centred around problems, particularly challenges of the 21st century, or what Honey et al. (2014) saw as opportunities for students to 'use knowledge and skills from multiple disciplines' on problem-based tasks that involve 'complex phenomena or situations' (p. 52)—the very idea that underpins integrated STEM.

Asking questions, challenging assumptions, dissecting problems into their component parts, applying disciplinary knowledge and skills to devise appropriate solutions to problems, testing the proposed solutions, and using evidence to refine solutions have become commonplace in integrated STEM classrooms. The inquiry processes involved in integrated STEM problem solving resemble that of science and mathematical inquiry and encompass processes that call for student (1) explanation of phenomena/issues/problems scientifically; (2) modelling of issues scientifically or mathematically; (3) construction and evaluation of designs for scientific inquiry, critical interpretation of scientific data and evidence; and (3) research, evaluation, and use of scientific and mathematical information for decision making. The inquiry processes for science and mathematics, and integrated STEM inquiry are similar yet different epistemically, particularly in their focus on relevance to real-world problems. Specifically, integrated STEM inquiry is usually characterised by the application of disciplinary knowledge and skills to generate and test plausible solutions. As illustrated by Bybee (2019) in his model for designing a STEM unit of work, students start with a problem situation and reach out to the knowledge and skills of science, technology, engineering, and mathematics to clarify the problem and generate solutions. He proposed that to solve a STEM problem, the 5Es (Engage, Explore, Explain, Elaborate, and Evaluate) are used throughout the whole process to gather, reason, and communicate information.

With the momentum for STEM problem solving gaining traction, and a call for greater emphasis on inquiry in STEM learning (Tan, Ong, Ng, & Tan, 2023), there is greater urgency for science, mathematics, technology, and engineering teachers to develop knowledge and skills on ways of integrating knowledge, epistemic practices, and norms of different disciplines to facilitate integrated STEM inquiry more productively. As such, educators seek to understand research results on STEM inquiry in order to incorporate evidence-informed best practices into their repertoire. Science teachers, for example, want to find out how inquiry-based learning in science can be adapted for integrated STEM problem solving. Similarly, teachers engaged with STEM problem solving face the concern of choosing suitable strategies to scaffold students' engagement and their understanding of the STEM problem-solving process. Students and parents may also be interested in how assessment practices in STEM problem solving differ from those of science and mathematics and if they should handle STEM assessment differently. In general, there is great interest in understanding how the introduction of integrated STEM inquiry in classrooms would change current science and mathematics teaching and learning behaviours.

Despite interest and best efforts, the process of developing competencies related to integrated STEM teaching and learning is challenging. Teachers must navigate a multitude of research evidence and recommendations from different theoretical stances and educational contexts before connecting the myriad of ideas to professional practices in their classrooms! The knowledge base related to integrated STEM teaching and learning is vast. Both the sensemaking and enactment processes are often time-consuming and confusing. Feedback from different communities of teachers and curriculum planners suggests a need for resources that connect the multitude of rich research findings, theoretical arguments, and practical wisdom. Teachers are eager to embrace promising ideas of integrated STEM inquiry but lack the self-efficacy to balance ensuring the fidelity of integrated STEM with the adaption to the specific demands of their local contexts.

Understanding the uncertainties faced by teachers and the educators of STEM teachers formed the motivation behind this book. This book hence aims to (1) distil the key research findings relevant to the planning and implementation of meaningful integrated STEM inquiry lessons and (2) connect these research findings to actual implementations of integrated STEM inquiry in the classrooms in the form of vignettes. Weaving theoretical frameworks and research findings to make sense of integrated STEM classroom practices is a complex task, and we hope that the examples provided in this book serve as meaningful starting points for professional conversations. We have been asked several times to clarify and describe more explicitly the intended audience of this book so that we can focus either on the theoretical ideas of integrated STEM or the events in classrooms. The artificial dichotomy and decoupling of theoretical ideas and classroom practices is potentially problematic as they separate theory from practice and in the longer term, practitioners might find theoretical ideas difficult to understand and irrelevant to classroom practices. Instead, we have intentionally ensured that the ideas presented in this book lie at the nexus of theory and practice. Since this book focuses on how integrated STEM learning can be interpreted from theoretical ideas of classroom practices and productive disciplinary engagements, we have deliberately paired classroom vignettes with detailed descriptions of theoretical ideas from various integrated STEM frameworks and lesson plans.

To ensure that the research and practice ideas presented in this book are accessible to both teachers of integrated STEM and STEM education researchers, we adopt a writing style that sits at the threshold of academic writing and narrative forms: a more academic slant when reporting research evidence and theoretical ideas and a more narrative and causal tone when describing aspects of teaching practice. We hope that this book would be a starting point and an easy reference for teachers who are beginning their integrated STEM journey. For STEM educators and researchers, the ideas presented in this book could form the basis of your courses for teachers or agendas for further research.

There are various ways to use this book. Each chapter focuses on a specific idea related to integrated STEM inquiry. Chapter 1 discusses the challenges faced by the world today as the motivation behind re-framing disciplinary siloed learning to a transdisciplinary form of problem solving. Here, we unpack some possible conceptions of integrated STEM inquiry held by teachers, distil the features of scientific and mathematical inquiry, and compare them to those of integrated STEM inquiry to persuade readers to reflect on their own conceptions. To assist in applying our understanding of integrated STEM inquiry to plan integrated STEM learning experiences, Chapter 2 presents the principles of integrated STEM learning from the perspective of the Next Generation Science Standards and the STEM Quartet instructional framework. We describe in detail three variants of the STEM Quartet as different ways to approach and plan for STEM inquiry. The value of this chapter lies in understanding the key pedagogical affordances behind each instructional framework to increase a teacher's self-efficacy in choosing the framework used to plan integrated STEM inquiry learning experiences. For teacher educators, ideas of the different instructional frameworks can be compared during courses to teach preservice and in-service teachers how the different frameworks can be applied to create meaningful learning experiences for learners with different learning profiles and learning needs.

Chapter 3 considers the common factors that teachers deliberate on in their choice of problems/issues presented to students and the strategies chosen to facilitate learning. Decisions related to the nature of the problems, relatedness to students' experiences, types and levels of disciplinary knowledge, strategies and tools for obtaining feedback, and levels of inquiry are all crucial considerations for successful enactment of STEM lessons. In Chapter 4, we delve into an area of great interest and concern among practitioners of integrated STEM inquiry—assessing students' progress and learning. In this chapter, we provide suggestions to answer the questions—how do we know that students have learnt? And how do we know what they have learnt? We suggest ways in which teachers can assess either or both the process as well as the final outcomes of engagement in integrated STEM inquiry. Teachers (both preservice and in-service) would find the examples described in this chapter familiar and easily applicable to their professional practice. For STEM education researchers, some of the assessment ideas (such as stealth assessment in STEM) could lead to new research agendas since these modes of assessment are relatively novel and could benefit from refinement, given more research evidence. In Chapter 5, we explore variations that are possible when planning and enacted integrated STEM inquiry across different educational contexts and conditions. Finally, Chapter 6 to 10, we use rich descriptions of vignettes and explanations to interpret the interactions between teachers, students, resources, and their environment to understand how integrated STEM inquiry is actualised. Especially, we highlight the connections between the theoretical ideas of STEM learning and practical enactment in classrooms.

References

Bybee, R. W. (2019). Using the BSCS 5E instructional model to introduce STEM disciplines. *Science & Children*, 56(6), 8–12.

Honey, M., Pearson, G., & Schweingruber, H. (2014). *STEM Integration in K-12 Education: Status, Prospects, and an Agenda for Research*. The National Academies Press.

Tan, A. L., Ong, Y. S., Ng, Y. S., & Tan, J. H. J. (2023). STEM problem solving: Inquiry, concepts, and reasoning. *Science & Education*, 32(2), 381–397.

Acknowledgements

The authors would like to acknowledge Temasek Foundation for funding this project under Temasek Foundation—STEM Programme in Thailand.

We are also thankful to teachers and students who participated in this project. We would like to express our gratitude for the collaboration and friendship of our partners, especially Dr Kanchulee Panyain, Miss Pondwalee Janduang (Ying), and Miss Nattakun Somkaew (Mint) from the Office of Basic Education Commission (OBEC), Thailand.

We would also like to thank our colleagues, Asst/P Park JoonHyeong, Dr Tan Lik Tong, Dr Johannah Soo, Dr Michael Tan, and A/P Lee Yew Jin, from meriSTEM@NIE for assisting with classroom observations. We are indebted to Professor Lyn English for taking the time to go through our manuscript and offer advice on how to improve our ideas. Finally, we are grateful to Dr Melissa Neo, who assisted with proofreading our drafts and Mr Ng Yu Heng and Mr Isaiah Leong for assisting with checking the index.

Any opinions, findings, conclusions, or recommendations expressed in this material are those of the author(s) and do not necessarily reflect the views of the NIE. NTU IRB Approval was obtained prior to data collection (IRB-2020-02-036).

1 Making sense of STEM inquiry

Reasons, inquiry, and problem solving

We live with a constant variety of problems. What problems are you currently living with? What do people care about in modern society? Let us look at the following news headlines from around the world from late 2022 and early 2023.

> European weather: Winter heat records smashed all over continent.
> (BBC, January 3, 2023)

> India's farmers tap technology to boost crop production amid unpredictable weather.
> (CNA, November 17, 2022)

> New variant XBB.1.5 is 'most transmissible' yet, could fuel COVID wave.
> (The Washington Post, January 8, 2023)

The above issues are problems/challenges that people around the world have been interested in and have fiercely dealt with in the 21st century. The first news item dealt with the problem of unpredictable weather in European cities believed to be caused by climate change, which, of course, is not just a European problem but a global one. The second news item similarly reported on changes in weather patterns and how people in India responded to the unpredictable weather with technology. The third item described a novel coronavirus variant that was emerging and quickly spreading in the United States as the pandemic continued to wreak havoc around the world.

Problems such as the unpredictable weather reported by the BBC will become increasingly difficult to predict and more challenging to solve. We live in an era with an urgent need for experts and talents who can sensitively understand these complicated issues and make rational and informed decisions.

Common to the three news headlines is that all of them pose complex, extended, and persistent issues that cannot be solved with knowledge from just a single discipline. Rather, these issues and problems require a multitude of individuals with different knowledge, skill sets, and innovative ideas to come together in teams to collaborate, debate, and test solutions. Increasingly, individuals within work teams that specialise in one specific discipline now need to have some knowledge of other disciplines, such as information communication and science technology, since the influence of science, technology, and engineering is becoming increasingly complex and multi-dimensional. The constituents of working teams are also becoming increasingly diverse and inclusive, consisting not just of people with different domain expertise but also people from different sociocultural backgrounds, age groups, and interests. As we move from working individually to working in diverse teams, competencies such as collaboration, communication, ability to reflect, and inter- and intra-personal awareness become more and more important. These competencies, although described with different emphases, are widely recognised and accepted across different groups as essential for education and the workplace in the 21st century (OECD, 2005, Partnership for 21st Century Skills, 2006; Metiri Group and NCREL, 2003; American Association of Colleges and Universities, 2007). How can we build these expert teams of multiple talents?

In response to the changes and demands of this era, science and mathematics education needs to be reformed to ensure relevance to the challenges faced by the world. Integrating ideas and practices of engineering and technology into science and mathematics learning is seen as a way for science and mathematics learning to stay relevant to students' lives in the 21st century. Consequently, integrated STEM education has gained popularity and has found its way into school curricula in different education systems across the world. Numerous STEM curricula with different characteristics have been developed and implemented over the past few decades (Korea Foundation for the Advancement of Science and Creativity, 2016; NAE & NRC, 2009; NRC, 2011; Stohlmann et al., 2012; Teo & Choy, 2021). To begin our exploration of integrated STEM learning, let us examine your personal conception of STEM education.

1.1 Exploring teachers' conception of STEM learning

What do you think STEM is? Many of you have already heard of or are familiar with STEM, as STEM has been a hot issue in education for the past two decades. How then would you depict the relationship and interactions between the four STEM disciplines? The models in Figure 1.1

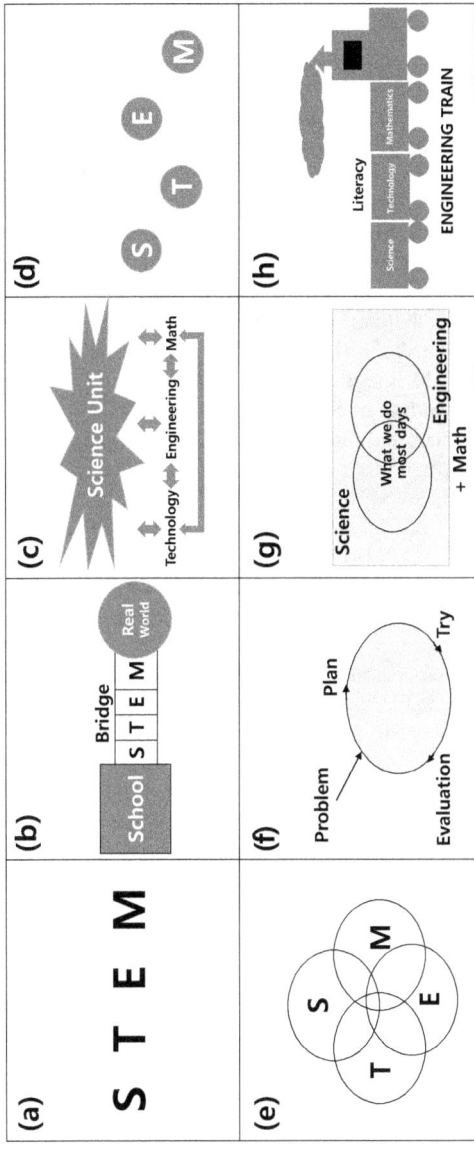

Figure 1.1 Teachers' conceptual models of STEM education (a) STEM as an acronym (b) real-world problem solving as context (c) science as context (d) science, technology, engineering, and mathematics as separate disciplines (e) integrated disciplines (f) engineering design process as context (g) science and engineering design process as context (h) engineering as context (Ring et al., 2017).

(Ring et al., 2017) show teachers' diverse conceptual understanding of STEM using visual representations. Your own understanding of STEM may resemble those presented in these models, or they may be different.

Let us examine the commonalities in teachers' perceptions of STEM, as depicted in Figure 1.1. First, most figures have the elements S, T, E, and M, representing each STEM discipline. We can see the common perception that science, technology, engineering, and mathematics are essential elements in STEM education. However, there are differences in how the disciplines of science, technology, engineering, and mathematics can be integrated to form integrated STEM. Some teachers perceive that a specific discipline, such as science or engineering, can form the context (Models C and H) of a learning experience; others believe that real-life problems or problem-solving processes can be a key aspect of a STEM lesson (Models B and F). In Models F and H, engineering is introduced into STEM learning primarily as the context where the problem is located. Which model of STEM integration appeals to you and is a closer reflection of your understanding of STEM?

The teachers' conceptual models of STEM shown in Figure 1.1 were formed differently depending on factors such as the teachers' specialisation majors, teaching experiences, or teaching and curricular contexts (Dare et al., 2019). Similarly, the teachers' STEM conceptual models, ideas, and curricula could also be interpreted and implemented in multiple pathways appropriate for various learner profiles, educational environments, sociocultural contexts, and broad STEM education goals (Teo et al., 2021).

However, even when creating multiple pathways for STEM education, it is essential to reflect the fundamental philosophy of STEM by considering how multiple disciplines can be effectively integrated. Primarily, in STEM education, it is important to create deep connections between disciplines to achieve meaningful integration. Which model (A to H) in Figure 1.1 do you think is a meaningful way of integrating the four disciplines of STEM? To answer this question, we first need to consider the following questions: What are the key disciplinary features defining each discipline? What are the similarities and differences between the four disciplines? What should be noted when blurring the boundaries between disciplines and integrating them? In the next section, we will discuss the characteristics and differences between science, technology, engineering, and mathematics and introduce educational perspectives to blur and integrate the boundaries between disciplines.

1.2 Understanding STEM disciplinary features

Given that every discipline has its unique disciplinary goals (cognitive, epistemological, and social), values, and ways to inquire and generate knowledge which are stable entities (Becher, 1994), creating deep connections between the four STEM disciplines is not a straightforward task. This is because the integration across the STEM disciplines requires a fundamental understanding of the differing characteristics of each discipline (Tan et al., 2019). The epistemic goals pursued by the STEM disciplines and their disciplinary practices can be summarised as shown in Table 1.1.

As the goals of each discipline differ, the disciplinary practices are also different. For instance, the primary goal of science is to inquire and explain how the natural world works while engineering aims to establish a body of knowledge largely related to man-made systems through the process of problem solving. Consequently, science as a discipline values the practice of inquiry and empirical evidence while engineering practices focus on the application of science and mathematics knowledge to overcome constraints and optimise the solutions for problems. Further, if we compare the disciplinary practices of science and mathematics, we realise that while empirical evidence is fundamental to the practice of science, logical arguments form the foundation of mathematical practices. Given

Table 1.1 Goals and disciplinary practices in the STEM disciplines (Bybee, 2019)

Disciplines	Goals	Disciplinary practices
Science	Inquiring and explaining the mechanism of *phenomena in the natural world*, including the laws of nature associated with physics, chemistry, biology, Earth, and space sciences	Science generates knowledge through the process of *science inquiry* based on *empirical evidence* to support or change claims
Technology	Establishing the system of people, organisations, knowledge, processes, and devices that go into creating and operating *technological artifacts*, as well as the artifacts themselves	Technology creates *a product of science and engineering* to satisfy the wants and needs of humans
Engineering	Establishing both a body of knowledge—about *the design and creation of human-made products*—and a process for solving problems	Engineering *utilises concepts in science and mathematics* as well as technological tools
Mathematics	Inquiring, identifying, and explaining *patterns and relationships* among quantities, numbers, space, and other abstract structures	Mathematics generates knowledge and its claims are warranted through *logical arguments* based on foundational assumptions. The logical arguments themselves are part of mathematics along with the claims

the differences in goals and disciplinary practices, it is evident that integrating the disciplines in practice can be tricky. An experienced chemistry teacher who was seeking to understand integrated STEM once asked if integrated STEM resembles a mixture or a compound! Indeed, this question highlights the complexities in establishing the integrative mechanisms for STEM. If integrated STEM is a mixture, it would adopt epistemic goals, norms, and practices of its component disciplines, although the proportion of each component discipline could vary. If integrated STEM is a compound, the epistemic goals, norms, and practices of the resultant discipline would be entirely different from its component disciplines. When referencing the teachers' conceptual models of integrated STEM shown in Figure 1.1, aside from Model E, all the other models presented integrated STEM as a 'mixture' of different disciplines put together to achieve the task of solving a problem or understanding a specific phenomenon. If integrated STEM learning is indeed a 'mixture' of four disciplines, the question to consider is 'Which disciplines should be harnessed, in what proportion, and how can the knowledge and practices of the disciplines be connected to one another?'

To better understand how epistemic goals, norms, and practices can be aligned in integrated STEM, we first seek to define integrated STEM so that we can identify integrated STEM practices when we encounter them. The way integrated STEM is conceptualised should also lead us to understand the way the four disciplines are interconnected. Integrated STEM has been defined as 'approaches that explore teaching and learning between/among any two or more of the STEM subject areas, and/or between a STEM subject and one or more other school subjects' (Sanders, 2009, p. 21). The important ideas emphasised by Sanders (2009) are 'approaches' and the connection between 'two or more disciplines' or 'one or more school subjects'. At this juncture, several questions arise:

(1) What do 'approaches' refer to?
(2) Which aspect of each discipline should be integrated?
(3) How can inquiry be coupled with problem solving?
(4) How can integration take place?

To answer the questions, we first explore the features of science and mathematical inquiry and move on to examine how integration across the disciplines can be achieved.

1.2.1 Inquiry as an educational approach has a long history

Inquiry is a universal educational approach that has been acknowledged in various fields of education. The fundamental spirit of inquiry can be traced historically to Socratic inquiry in ancient Greece (Friesen & Scott,

2013). Socratic inquiry encompasses asking probing questions to clarify the basic assumptions or the logical arguments that support or challenge truth claims (Ross, 2003). Although this can be viewed as a philosophy of truth-seeking, it is not easy to translate this into a series of educational approaches (Friesen & Scott, 2013).

Inquiry as a pedagogical approach was established in the modern era (Friesen & Scott, 2013). John Dewey (1910, 1938), an American philosopher and educational reformer, established an inquiry-based approach in science learning. Dewey considered the process of inquiry to involve 'sensing perplexing situations, clarifying the problem, formulating a tentative hypothesis, testing the hypothesis, revising the solution with rigorous tests, and acting on the solution' (Barrow, 2006, p. 266). He argued that students must be taught to pose problems related to their experience and augment emerging understanding with their knowledge (Dewey, 1938).

To many of us living in the modern era, the philosophy of posing questions to satisfy our individual curiosity may seem natural. However, from a pedagogical point of view, in the early 20th century, when systematic education system was not fully established to meet the needs of every student, active inquiry was a very innovative approach (Cohen & Hanagan, 1991; Gillard, 2011). During that period, Dewey's philosophy of emphasising active learning was deeply anchored in students' lives and still shares a fundamental spirit with several recent teaching approaches, including those in STEM education. Thus, this innovative inquiry-based approach has been considered a desirable basis for learning, not only in mathematics and science (e.g., American Association for the Advancement of Science [AAAS], 1993; Anderson, 2002; NRC, 1996; National Council of Teachers of Mathematics [NCTM], 2000) but also in other subjects such as languages and social studies (Grant, Swan & Lee, 2017; Short et al., 1996).

For the past few decades, educators have applied the inquiry-based approach to learning activities in several subjects with a focus on commonalities of student learning (Grant et al., 2017). For example, inquiry involves two interrelated processes: developing ideas related to the conceptual subject matter and engaging in sensory-related activities (Poon et al., 2012). This inquiry process allows students to broaden their personal and social understanding of the world by using various experiences and knowledge from each discipline, such as mathematics, science, language, and arts, as active learners (Short et al., 1996).

Given its long history of application in many subjects, the inquiry-based approach offers a variety of possibilities as an overarching process in STEM education that emphasises integration across multiple disciplines (Anderson & Li, 2020; Tan et al., 2019). Based on an understanding that inquiry can be applied across multiple disciplines as a pedagogical approach, the next section will examine the affordances of inquiry-based learning practically and more specifically through inquiry in science education.

1.2.2 Science inquiry as an epistemic practice of the science discipline

Schwab (1962), a science educator, curriculum scholar, and one of the founders of the concept of science inquiry, argued that teaching science as the rhetoric of conclusions should be avoided. This is because delivering scientific knowledge as the final truth can give students a distorted perception of science, since the process of forming scientific knowledge involves questioning, collecting empirical data, making sense of data, and forming plausible explanations. Further, scientists engage with debates over ideas, theories, and evidence over prolonged periods of time before 'finalising' a proposed scientific theory or concept. Moving away from teaching facts about science towards more active engagement in questioning, experimenting, and forming explanations was advocated by Dewey as early as 1938. To date, most science educators have considered it essential to provide practical and authentic science inquiry that can promote students' active thinking when learning.

As the idea of inquiry gained acceptance, Schwab (1962) argued that inquiry in science education simultaneously includes two conceptual meanings: (1) science as inquiry and (2) teaching-learning as inquiry. The concept of science as inquiry focuses more on inquiry as the method of generating scientific knowledge. That is, through the process of science inquiry, students experience the essence of the epistemic practices of science in science classrooms (Duschl & Grandy, 2008). On the other hand, the concept of teaching-learning as inquiry centres on how to teach and learn science. As discussed, this concept means inquiry as a general teaching pedagogy can be applied in many disciplines. In this book, our discussions cover both science as inquiry and teaching-learning as inquiry since inquiry forms one of the central ideas of our arguments.

The discussion of 'science as inquiry' has a history as long as 'teaching and learning as inquiry'. Given that scientific knowledge is generated through scientific inquiry, as shown in Table 1.1, students need to experience the epistemic practices of constructing knowledge through science inquiry to appreciate the nature of science. The need for students to actively engage in inquiry to learn does not apply only to science but also to other disciplines that rely on iterative inquiry cycles to construct productive and valid ideas (Moore et al., 2015). Specifically, inquiry-based learning creates opportunities for students to construct and develop disciplinary knowledge through inquiry as an epistemic practice.

Science educators have had many discussions about what meaningful science inquiry is and how to practice it. For the past few decades, the concept of science inquiry has gradually changed in its meaning and in its defining characteristics. Concretely, the perceptions of science inquiry have altered from emphasising content and skills to a form that gives priority to evidence-based reasoning, social context, and interaction (Duschl & Grandy, 2008).

While science inquiry includes processes such as learners asking scientifically oriented questions, gathering evidence, formulating explanations, evaluating explanations in light of new evidence, and justifying proposed explanations (NRC, 2000), these inquiry processes have extended beyond science inquiry to the more holistic construct of 'science practice'. Science and engineering practices presented in *Next Generation Science Standards* [NGSS] (NGSS Lead States, 2013) show this change in a concerted manner. The NGSS practices involve epistemic practices in generating scientific knowledge, such as forming questions that can be answered by inquiry, analysing, and interpreting data in ways accepted within disciplines, and constructing explanations, models, or theories based on empirical data. Interestingly, there are some science and engineering practices that overlap. For example, 'constructing and using models and mathematical and computational thinking' are presented in both science and engineering. Practices that are common to both science and engineering can become intersections that can be applied in integrated STEM.

1.2.3 Mathematical inquiry as a disciplinary practice

Mathematics is often seen as an accessory to science and the other disciplines in integrated STEM tasks even though it is one of the key disciplines in STEM. The role of mathematics is relegated to 'mainly operative work', 'a means to an end', with the 'problem-solving part assigned to the other disciplines' (Just & Siller, 2022, p. 16). However, mathematics can certainly play a more prominent role in integrated STEM by considering how the disciplinary practices of mathematics can be strengthened. This begs the question: What does it mean to do mathematics, especially in school contexts?

In short, doing mathematics in secondary and tertiary levels can be perceived as adopting a set of actions, ways of working or habits of mind, and a productive mindset, similar to those employed by mathematicians confronted with a complex problem (Yeo & Choy, 2023). At the heart of doing mathematics lies the notion of mathematical thinking—the act of specialising, conjecturing, justifying, and generalising (Mason et al., 1985)—as well as the ability to pose problems for investigation (Yeo, 2017; Yeo & Yeap, 2010), and solve them (Schoenfeld, 1992). To specialise means to look at specific examples or cases, to look for patterns and structures, so that one can begin to generalise or say something about the underlying patterns or structures, which can thereafter be applied or extended to other cases. The inquiry process then goes beyond specialising and generalising to conjecturing—making an intelligent guess of what the pattern or rule might be. But truth in mathematics is not merely established by thinking about specific cases and generalising them to form conjectures. Instead, it involves constructing a series of logical statements that either lead towards a proof or an argument demonstrating that a conjecture is true or a counterexample to prove that a conjecture is false. These processes are invoked when mathematicians pose and solve problems. These actions, while similar to science inquiry in some ways, differ in the weight given to logical reasoning over observation making.

In particular, the way of working in mathematics is underpinned by two intertwined forms of reasoning: inductive and deductive reasoning. On one hand, inductive reasoning can be conceived as the process of developing conclusions, arguments, explanations, judgements, and inferences from observing mathematical objects such as ideas, concepts, and representations (Brodie, 2010). On the other hand, deductive reasoning involves the process of formulating a formal proof, which includes generating mathematical arguments to explain and justify mathematical observations based on a set of pre-established premises. Henningsen and Stein (1997) see this ability of making sense of and reasoning about mathematical ideas in 'flexible ways' as part of 'mathematical dispositions' (p. 525).

These mathematical dispositions can be seen as productive mindsets necessary for learners to adopt in the face of a problem. These mindsets could include a willingness to embrace new challenges, persisting through struggles, and thinking flexibly to see the problem from different perspectives (Costa et al., 2022). These productive mathematical mindsets are closely associated with 'methods' of generating ideas when students work on mathematical problems or investigative tasks (Isoda & Katagiri, 2012, p. 49). These different aspects of doing mathematics thus position mathematical problems and investigative tasks at the centre of a mathematics classroom focused on fostering such disciplinary thinking.

The emphasis on problems and investigative tasks is largely aligned with two key orientations to the learning of mathematics, as advocated in recent revisions of mathematics curricula in Singapore: learning mathematics as a tool to solve problems and learning mathematics as a discipline. To learn mathematics as a discipline means to learn how to think and do mathematics like mathematicians. With the aim of developing mathematical thinking in our learners, there is a need to develop their capabilities in understanding the purpose of disciplinary expertise, its inquiry methods, and its forms of communication, in addition to an essential base of knowledge (Mansilla & Gardner, 2008). This multi-dimensional

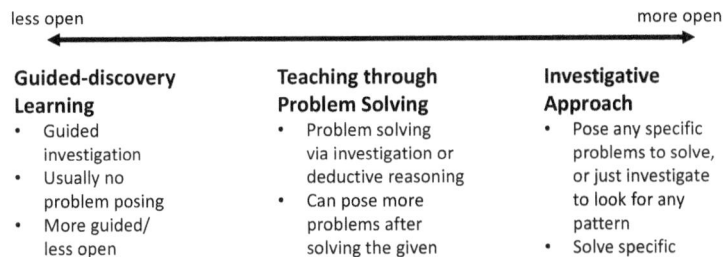

Figure 1.2 Guided discovery to investigative approach (Yeo & Choy, 2023).

view of mathematics was also echoed by Devlin (2000) who argued that students should experience all four faces of mathematics—mathematics as 'computation, formal reasoning, and problem solving'; 'a way of knowing'; 'a creative medium', and 'applications' (p. 16)—to develop the knowledge and competencies necessary for living 'full and active lives' in an ever-changing world (p. 17).

To this end, teachers can think about the use of guided-discovery learning tasks, teaching through problem solving, and open investigative tasks in their teaching (Yeo & Choy, 2023). Some of the key ideas related to these tasks are summarised in Figure 1.2.

These different tasks provide a means for teachers to emphasise mathematics as a 'way of knowing' and as a 'creative medium' beyond the usual notions of mathematics as 'computation, formal reasoning, and problem solving' and, to a limited extent, 'mathematics as applications' (Devlin, 2000, p. 18). Similarly, in the context of integrated STEM, the use of complex and persistent problems in learning can provide new possibilities for students to engage with mathematics beyond computation and, instead, engage with the disciplinary practices of a mathematician as they view these problems through the lens of mathematics.

1.2.4 Towards STEM inquiry: Conceptual and pedagogical perspectives

The inquiry-based approach has also been applied widely in STEM and is one of the most popular pedagogies in STEM learning (Anderson & Li, 2020; Moore et al., 2015). Focusing on investigative problem solving as a similar epistemic practice in mathematical and scientific inquiry, we

12 Making sense of STEM inquiry

examine the conceptual and pedagogical basis of integrated STEM inquiry by making comparisons to general inquiry. In this section, we explore the conceptual validations and pedagogical possibilities of STEM inquiry and conceptualise integrated STEM inquiry as an overarching process.

From a conceptual perspective, our earlier discussions delved mainly into the crucial roles played by inquiry when developing disciplinary knowledge in mathematical and scientific disciplines. Indeed, it has also been reported that the epistemic underpinning of integrated STEM is similar to those of mathematics, science, and engineering. The inquiry that occurs in integrated STEM can be characterised as understanding phenomena, logical reasoning based on related background knowledge, verifying and evaluating reasoning, and drawing constructive and valid ideas (Moore et al., 2015). In this vein, STEM knowledge can be described as 'theories, laws and explanations that underpin the outcomes of STEM inquiry' (Park et al., 2020, p. 141), underscoring the epistemic nature of STEM. Hence, the epistemic nature of inquiry is similar not only across science and mathematics but also in general inquiry used in sense making in daily life (Park et al., 2020).

Disciplinary inquiry and general inquiry are positioned along a continuum with general or common-sense inquiry differing from disciplinary inquiry in terms of logic. According to Dewey (1938), the core of the inquiry processes is similar for both general inquiry and science/mathematical inquiry, even though the object being investigated may differ. Figure 1.3 compares the similarities and differences between general

	General Inquiry	Science Inquiry	Mathematical Inquiry		STEM Inquiry
Differences	• Serves more practical purposes • Inquiries about immediate problems in everyday life	• Inquiries about subjects that are less familiar in everyday life, such as spintronics or photosynthesis	• Focuses on examining problems for patterns and structures and using logical reasoning for justification	→	• Serves both practical and domain disciplinary purposes • Epistemic practices of generating STEM knowledge • Enables students to learn disciplinary knowledge through inquiry about authentic problems in real life • Implemented using various inquiry-based instructional models • Values reasoning based on both conceptual and practical success factor systems
Similarities	• Epistemic practices of inquiry encompass formulating questions, obtaining data, and finding answers or reasonings based on evidence • Constructing and developing knowledge or crafting conjectures through inquiry				

Figure 1.3 Conceptualisations of STEM inquiry based on general inquiry, science, and mathematical inquiry.

inquiry, science, and mathematical inquiry to help us establish the concepts of STEM inquiry.

General inquiry usually deals with immediate problems in everyday life, while mathematical and science inquiry is more likely to inquire about subjects that are more remote from familiar everyday life experiences, such as number theory, spintronics or photosynthesis (Tan et al., 2019). Similarly, general inquiry serves more practical purposes compared to mathematical or science inquiry. For instance, while general inquiry may be used to understand why a student is frequently late for school, science inquiry may be used to question the evidence available to explain the phenomenon of the rate-limiting step in traffic flow, while mathematical inquiry may be used to look for patterns in the clustering of buses to propose a logical model to predict traffic jams. However, much like mathematical and science inquiry, the process of general inquiry also involves formulating questions, obtaining relevant data, and finding answers or crafting conjectures based on evidence. However, the object of inquiry could be different. The similarities in epistemic practices between mathematical and science inquiry and those of general inquiry suggest possibilities for adaptation and integration of different disciplinary inquiry practices in integrated STEM inquiry. For instance, when presenting the context of a problem, general inquiry is likely to be useful in helping students understand the sociocultural context of the problem. Students and teachers can discuss questions such as 'How frequent is the problem observed?', 'How many people are affected by the problem?', 'What are some common complaints heard?' However, just focusing on general inquiry may not allow students to fully appreciate the complexity of the problem. For instance, understanding that the problem of a traffic jam occurs every day and that thousands of inhabitants are affected by traffic jams will not make students appreciate the possible causes of traffic jams. Students can engage with mathematical inquiry by conjecturing about the causes of traffic jams and design testable investigations to collect data. They can also pose more technical questions related to science such as 'How is the average velocity of vehicles travelling on expressways calculated?' and 'How are traffic lighting systems connected and controlled?' From the examples illustrated, it is evident that integrated STEM inquiry builds upon both general, mathematical, and scientific inquiry and the collective benefits of using the three types of inquiry are greater than any one on its own.

Other than the epistemic practices of inquiry, disciplinary knowledge is also required for students to engage in meaningful STEM inquiry. Zimmerman (2000) conceptualised the idea of domain-specific knowledge and domain-general knowledge, which are similar in general, mathematical, and scientific inquiry. Zimmerman argued that learning in all disciplines must involve domain-specific knowledge and skills as reliance on domain-general knowledge and skills limits students' understanding of epistemic goals of the disciplines. For instance, if students are to learn to connect scientific evidence to theory, presenting them with an example of a misplaced object and asking them to determine how and why the objects are misplaced does not help students learn any scientific or mathematical knowledge. It also does not present opportunities for students to engage with the rigorous process of testing scientific knowledge or writing statements of logical reasoning. In other words, students would be unable to inquire scientifically or mathematically if presented with purely social, real-world, or political issues. As such, as integrated STEM educators we need to intentionally map and connect the forms of inquiry to the corresponding disciplinary epistemic goals of the various STEM disciplines.

From a practical perspective, the inquiry-based approach promotes students' enactive thinking, participation, and effective learning. Tan et al. (2019) explained the effectiveness of an inquiry-based approach in STEM in bridging the gaps between disciplinary knowledge (science and mathematics) and life experiences. Additionally, an inquiry-based approach can potentially address several pertinent issues in existing STEM practices. For example, research findings suggest that when students are immersed in engineering design or STEM problem solving, they often fail to learn mathematical or scientific content (Berland & Steingut, 2016). These problems arise because learning disciplinary content does not occur spontaneously in students' minds; it becomes secondary to the integrated STEM problem-solving learning experience (Kanter, 2010; Puntambekar & Kolodner, 2005). The connection between disciplinary knowledge and problem solving thus needs to be made intentionally and STEM inquiry offers a means for this to happen (see Zimmerman, 2000 described earlier). Students can be guided to learn the difference between reasoning that is based on a value system of success criteria in problem solving and reasoning that is based on a conceptual system centred on disciplinary knowledge and norms (Eshach, 2008). Through the integrated STEM inquiry process, students can be led to develop an understanding of disciplinary content in STEM by explicitly articulating the specific subject-matter knowledge chosen and applied to solve the problem. In this way, STEM inquiry can be a practical supplement to the missing parts in STEM education.

Focusing on the epistemic and practical possibilities of inquiry, scholars have developed and adapted many inquiry-based models or programmes for integrated STEM learning activities. As discussed earlier, one good example of an integrated STEM inquiry framework is the 5E model proposed by Bybee (2019). The 5E model allows students to 'engage' in problem-based situations, and 'explore' and 'explain' problem situations by applying knowledge and processes from the STEM disciplines. Students further 'elaborate' their ideas and understandings through additional activities and 'evaluate' them. The 5E model was originally developed specifically to promote and facilitate inquiry for science teaching as it captures the core of scientific inquiry well (Bybee, 2002). The applicability of the 5E model to STEM classes demonstrates the epistemic and pedagogical affordances of integrated STEM inquiry for effective STEM teaching and learning and its shared similarities to science inquiry.

In another example of STEM inquiry, Wang et al. (2020) developed an integrated STEM inquiry framework that considers the goals, inquiry design, implementation, and evaluation of lessons and applied it to a STEM programme for elementary students. In this programme, they investigated the learning opportunities provided by the STEM inquiry activities. In particular, the integrated STEM inquiry programme afforded students' construction of STEM disciplinary knowledge and skills.

Based on our discussions thus far, one can distil the characteristics of integrated STEM inquiry. First, integrated STEM inquiry is the epistemic practice of generating STEM knowledge. By STEM knowledge, we refer to disciplinary knowledge, understanding connections between disciplines, and the ability to apply specific and appropriate knowledge to problem solve. Second, STEM inquiry affords students' enactive thinking, participation, and effective learning. In particular, STEM inquiry bridges the gap between subject knowledge and experience, enabling students to learn disciplinary content effectively through and from authentic problems and their solutions. Third, STEM inquiry can be implemented and supported through various inquiry-based instructional models such as the 5E model (Bybee, 2019) and the integrated STEM inquiry framework proposed by Wang et al. (2020).

Let us now examine different ways to integrate the knowledge and practices of the different disciplines to create a meaningful STEM inquiry experience.

1.3 Understanding integration models

This section explores different integrative mechanisms used to achieve coherent integrated STEM learning. One of the most popular ideas on levels of integration suggested by Vasquez et al. (2013) is summarised in Table 1.2.

The examples described in Chapters 6 to 10 of this book are transdisciplinary as they involve students working with real-world problems and applying knowledge and skills from different disciplines. The lessons were designed using the STEM Quartet instructional framework (Tan et al., 2019), which has real-world problems or solutions at its core. While we aspire to engage students with transdisciplinary learning since this is the highest level of integration, we have to be mindful that before students can engage with solving real-world problems, teachers and students must spend time developing their disciplinary knowledge in disciplinary,

16 *Making sense of STEM inquiry*

Table 1.2 Different levels of integration

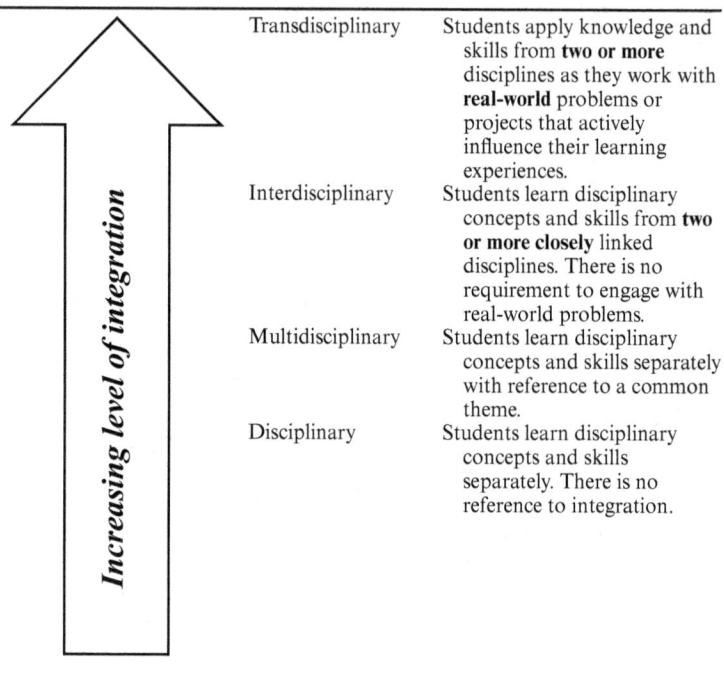

	Transdisciplinary	Students apply knowledge and skills from **two or more** disciplines as they work with **real-world** problems or projects that actively influence their learning experiences.
	Interdisciplinary	Students learn disciplinary concepts and skills from **two or more closely** linked disciplines. There is no requirement to engage with real-world problems.
	Multidisciplinary	Students learn disciplinary concepts and skills separately with reference to a common theme.
	Disciplinary	Students learn disciplinary concepts and skills separately. There is no reference to integration.

Source: Adapted from Vasquez et al., 2013.

multidisciplinary, or interdisciplinary ways. In other words, to work on transdisciplinary problems, students must first develop disciplinary understanding monodisciplinarily and develop their ability to make connections between the knowledge and skills from two or more disciplines through multidisciplinary or interdisciplinary learning experiences.

Besides Vasquez et al.'s. (2013) levels of integration, researchers have also proposed other models of integration with the intention of achieving greater combined synergy across disciplines. When planning for integration, it is meaningful to structure the relationships and roles of the disciplines effectively based on the unique features of the STEM disciplines. For example, Bybee (2019) presented a STEM model (Figure 1.4) in which students mentally reach out to the four disciplines to address problem situations. Kelley and Knowles (2016) on the other hand visualised the role of each discipline as one of four pulleys that lift the block of situated STEM learning (Figure 1.5). In 2019, Tan et al. (2019) conceptualised the connections between the STEM disciplines during the problem-solving

Making sense of STEM inquiry 17

Figure 1.4 Students bring in and integrate STEM disciplines with the problem situation (Bybee, 2019).

process. In this model (Figure 1.6), the differences in line thickness among the disciplines represent the different strengths of connections between the STEM disciplines and the problem.

As shown in Figures 1.4–1.6, STEM conceptual frameworks typically contain two overarching processes: inquiry and problem solving. First, science inquiry is one of the big pulleys in Kelley and Knowles's (2016) model, playing an important role in lifting-situated STEM learning. The 5E learning cycle, initially developed as a systematic model of inquiry instruction (Bybee, 1997), has been reinterpreted and widely used as a STEM instructional model (Bybee, 2019). This shows that the inquiry process can likewise be applied to STEM classes, albeit in a different form. Second, problems are central to the STEM models of Tan et al. (2019) and Bybee (2019). In particular, Tan et al. (2019) visualised the problem-solving process with a round outer frame (Figure 1.6), indicating that problem solving is an overarching process that occurs in dynamic relationships with various disciplines. The reinterpreted 5E model for STEM learning similarly features a problem situation at the centre and describes how students reach out to the knowledge of the four disciplines to solve the problem. In STEM learning, the two processes of inquiry and problem solving are intertwined and play critical roles in engaging learners to ask meaningful questions to make sense of the problem and gain better insights into issues before identifying the knowledge, technology, and skills required to generate a solution.

1.4 Role of problems and problem solving in STEM

1.4.1 Key points for establishing STEM problems and problem solving

Problem solving is another overarching process that plays a vital role in integrated STEM learning; it is an approach widely used in educational settings to bridge the gaps between disciplinary knowledge and life experiences. Learning through problem solving has been adapted and

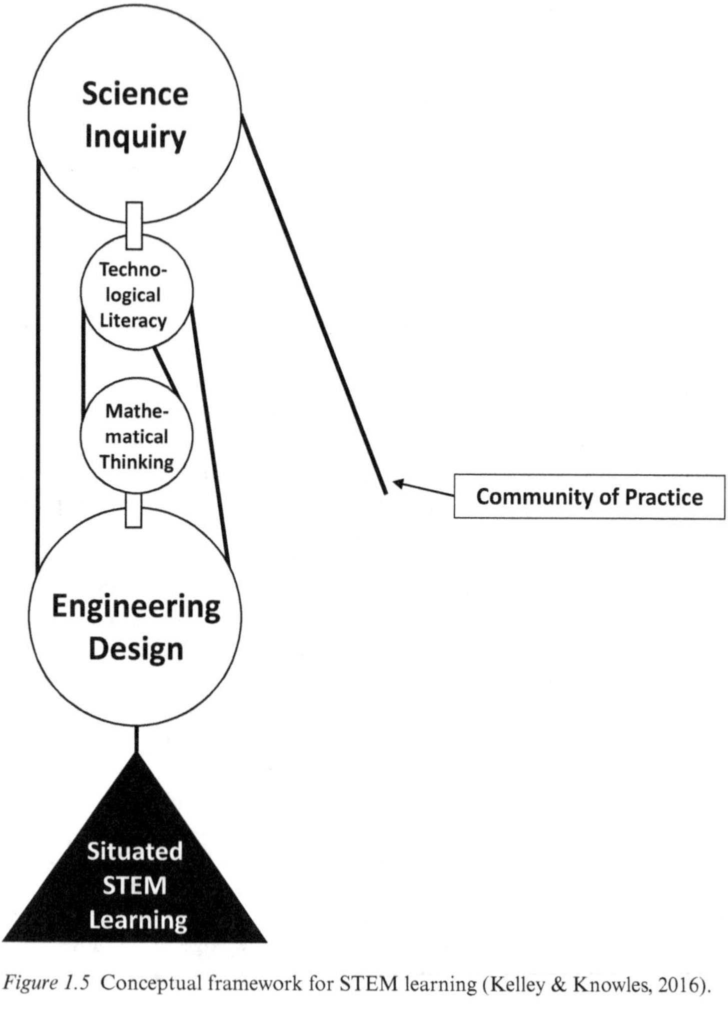

18 *Making sense of STEM inquiry*

Figure 1.5 Conceptual framework for STEM learning (Kelley & Knowles, 2016).

variously discussed in several fields of education such as mathematics, medical education, and engineering education (English, 2023; Kirsh, 2009; Delahunty et al., 2020). The same goes for STEM learning. In many integrated STEM instructional models, students usually start with authentic problems in real life (Banks & Barlex, 2014; Bybee, 2019; Madden et al., 2013; Tan et al., 2019) that require knowledge of either science, technology, engineering, or mathematics to work out solutions. The real-world problems presented align with transdisciplinary problems described by Vasquez et al. (2013) in Table 1.2. However, not all real-world

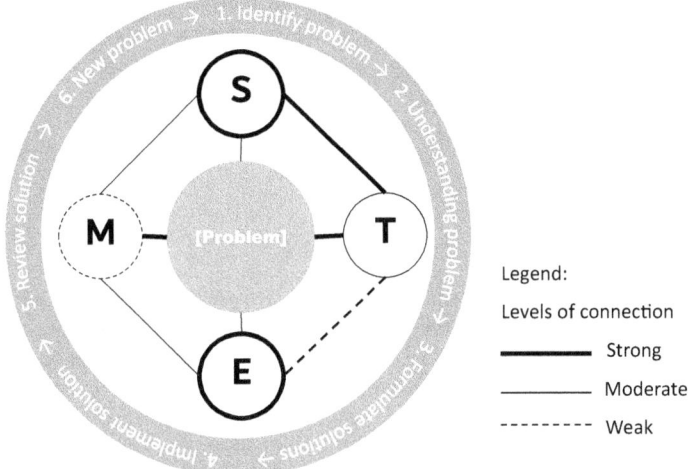

Figure 1.6 S-T-E-M Quartet instructional framework (Tan et al., 2019).

problems are STEM problems. For instance, efforts to promote religious harmony relate more so to a social problem than a STEM problem. As such, authentic real-world problems that require knowledge from STEM disciplines are defined as STEM problems in this book.

Some examples of transdisciplinary STEM problems we have observed and describe in the later chapters include (1) greasy wastewater polluting the school canteen, (2) health threats from excessive pigeon droppings in school, (3) sending water to remote mountainous villages, and (4) improving the work conditions of rubber tappers. STEM problems involve multifaceted features in terms of context, structure, and content. First, STEM problems are problems that can be dealt with from a realistic context in a novel and accessible manner (Teo et al., 2021). It is difficult to stimulate students' curiosity about problems that are too obvious or too huge in spite of their real-world contexts. So, there is a need to scaffold and scope the problem to an appropriate 'size'. For instance, in the problem of greasy wastewater polluting the school canteen, students conducted observational studies to understand the origins of the greasy water and where it flowed to. The teacher and students worked together to define the problem by delimiting the wastewater source to the school canteen rather than dealing with wastewater problems of the entire school or village. This helped to ensure that the problem was not too huge for the students to handle.

If an identified STEM problem is common in the lives of students, effort needs to be made to enable students to learn about aspects of the problem that are not always obvious, as familiarity might render the problem 'invisible' to the students. For instance, when presenting the problem of pigeon droppings, students can carry out an observational study of the changes in areas covered by pigeon droppings throughout the week. Data collected can help students problematise the familiar. As students encounter pigeon droppings every day, the droppings can form part of the students' school experience and become 'invisible' to them; students are hence unlikely to view the droppings as a health threat. As such, collecting data enables them to develop a more acute sense of the issues related to pigeon droppings.

Second, STEM problems are more ill-structured rather than linear (Tan et al., 2023a). Identifying the complex structure of a problem can promote students' creative and critical thinking (Nadelson & Seifert, 2017). In the problem of the lack of water supply in remote mountainous regions in Thailand, students are required to systematically identify the different factors that make up the problem and determine how the factors are connected to one another. Factors such as infrastructure, rainfall patterns, mountain terrain, economic viability, and accessibility to water are all issues that can be linked to the problem. Unlike linear problems that are typically solved by understanding the cause and effect, complex STEM problems require students to identify component parts, relationships, trade-offs, and limitations.

Third, STEM problems are problems dealing with multidisciplinary content (Rinke et al., 2016). When identifying and solving STEM problems, students need to be guided to connect multidisciplinary knowledge (Tan et al., 2023a). For instance, in the problem of poor work conditions for rubber tappers, students can apply knowledge of the effects of heat on human health, the biology of rubber trees, the optimal rate of tapping rubber, and of course, the motivation of human behaviour. The comprehensive range of knowledge and skills that are required for making sense of the problem and for generating a plausible solution go beyond science, mathematics or technology. Rather, it requires the combination of content, skills, and thinking from multiple disciplines to create a solution. To solve these complex problems, students require a wider knowledge base. English (2023) proposed a framework for solving STEM-based problems. In her framework, English considered critical thinking, critical mathematical modelling and philosophical inquiry, systems thinking, and design-based thinking as important processes and outcomes of problem solving.

With regard to the types of knowledge in multidisciplinary STEM problem solving, we refer to two types that differ depending on their epistemic nature: referent-centred and problem-centred knowledge (Bereiter, 1992). Referent-centred knowledge is organised around referents, such as the subject matter of textbooks. In contrast, problem-centred knowledge

Making sense of STEM inquiry 21

is organised around a problem, such as a real-life issue. Bereiter (1992) argued that school curricula tend to focus on referent-centred knowledge, which causes problems of inertness. Students need to be led to use problem-centred knowledge to solve problems in their STEM learning so that learning becomes active.

In this vein, there are two key considerations when designing STEM problems: relevance of problems to students' lives and the knowledge required to understand the problem. Upon establishing the novelty and challenges of a STEM problem, considerations can be made to assess how the problem can be presented to prioritise problem-centred knowledge. Problem-centred knowledge can serve as critical material and starting points for students to explore and solve problems. Thus, teachers may consider restructuring referent-centred knowledge (i.e., curricula knowledge) into problem-centred knowledge as and when appropriate to the problem. Practical ways highlighting how STEM problems and problem-centred knowledge can be connected will be dealt with in detail in Chapter 3. To further develop our understanding of the theoretical foundation of integrated STEM inquiry, the following sections will examine the possibilities of STEM inquiry and its pedagogical functions.

1.4.2 STEM inquiry as a bridge between STEM problems, problem solving, and STEM knowledge

What do you think promotes student learning during problem solving? Successful problem solving is characterised by evidence of students applying their prior experiences, knowledge, and familiar representations to generate new representations and reasoning patterns that can resolve tensions or ambiguities (Lester & Kehle, 2003). In other words, students develop the ability to connect problems lacking meaningful information with their personal relevant experiences and knowledge to model reasoning patterns or solutions, which eventually leads to meaningful learning.

In a dialectical manner, meaningful learning promotes students' self-efficacy in STEM problem solving, which leads to a better learning experience. Indeed, Tan et al. (2023a) analysed students' conversations during integrated STEM problem solving and reported that students make various inferences to develop solutions by combining their everyday experiences and knowledge. Let's take a closer look at Figure 1.7 as an example. In this integrated STEM activity, the students were presented with the problem of lack of land for farming despite the need to fulfil a minimum crop production of 120 tonnes of *Kailan* (a common leafy green vegetable in South-east Asia). While the students may not be familiar with farming methods, they have subject-matter knowledge about the process of photosynthesis, the conditions needed for photosynthesis, and the products for photosynthesis. As such, when the teacher introduced students to four existing vertical farming methods, students were readily able to identify the affordances of the vertical farming systems, their advantages, and their disadvantages. Subsequently, students could adapt and redesign current farming systems to create improved versions of current vertical farm systems. The students constructed a prototype of their proposed structures, tested, and presented their prototypes in class for peer critique (Tan et al., 2023b).

In STEM problem solving, students make decisions about the kinds of knowledge in the fields of science, engineering, technology, and mathematics that they can apply to generate solutions for solving STEM problems (Bybee, 2019). The knowledge output of integrated STEM problem solving can be new disciplinary knowledge, an improved solution, a novel solution, a new way to test an output, a better understanding of how to test solutions, or a changed perspective of how disciplinary knowledge and skills from different disciplines are connected. This reasoning of connecting and reconstructing knowledge in STEM problem solving is a crucial part of STEM inquiry (Tan et al., 2023a). In Figure 1.7, the four rectangles describe the disciplinary knowledge in each individual domain (vertical knowledge), and the arrows connecting these rectangles show how the disciplines are connected (horizontal connections). How these connections are made constitutes part of the process of integrated STEM inquiry.

In short, students develop an understanding of the relationship between problems, subject-matter knowledge, and disciplinary connections through engagement with the integrated STEM inquiry process during STEM problem solving. Students can learn by organising knowledge and experience to understand a problem (Bereiter, 1992). Furthermore, integrated STEM inquiry affords students opportunities to generate and restructure different ideas and explain specific design considerations, fostering higher-order thinking skills and competencies (Tan et al., 2023a). This is why we should pay attention to STEM inquiry as a bridge that connects problem solving, relevant subject-matter knowledge, and everyday experiences in STEM problem solving.

Furthermore, as shown in Figure 1.7, STEM inquiry can be a pedagogical lens that highlights the specific disciplinary knowledge and skills related to and required to make sense of a specific STEM problem, reveals how knowledge from different domains is connected, and illustrates how knowledge and experiences can be applied into meaningful solutions. Based on these theoretical discussions of STEM problems, problem solving, and STEM inquiry, the next chapter will examine practical ways for connecting and integrating knowledge from different disciplines through integrated STEM inquiry.

Making sense of STEM inquiry 23

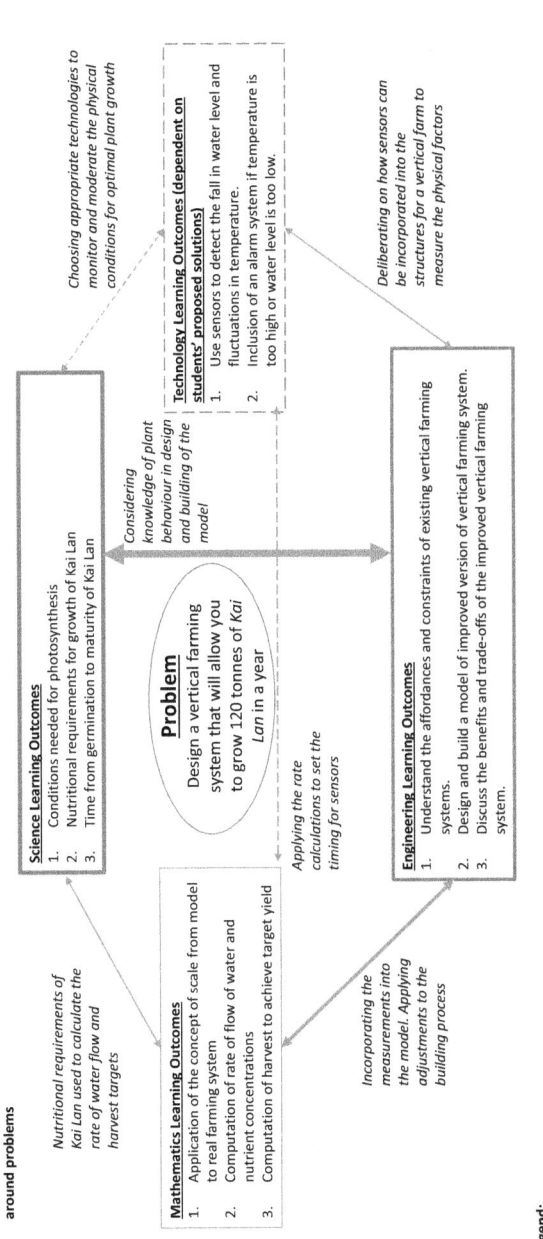

Figure 1.7 Connections across disciplines in an integrated STEM activity (Tan et al., 2023a).

Questions for Reflection

1. What are some conceptions of STEM learning that you or your peers have? Are they similar to the conceptions described by Ring et al. (2017) in Figure 1.1?
2. Examine the three instructional frameworks for integrated STEM learning described in Figures 1.4–1.6. What are the strengths and limitations of each framework? Which framework is most useful for your practice and why?
3. What are the similarities and differences between mathematical inquiry, science inquiry, and integrated STEM inquiry?

References

American Association for the Advancement of Science [AAAS]. (1993). *Benchmarks for science literacy*. Oxford University Press.

American Association of Colleges and Universities. (2007). *College learning for the new global century*. AACU.

Anderson, J., & Li, Y. (Eds.). (2020). Investigating the potential of integrated STEM education from an international perspective. In J. Anderson and Y. Li (Eds.). *Integrated approaches to STEM education* (pp. 1–12). Springer.

Anderson, R. D. (2002). Reforming science teaching: What research says about inquiry. *Journal of Science Teacher Education, 13*(1), 1–12.

Banks, F., & Barlex, D. (2014). *Teaching STEM in the secondary school: Helping teachers meet the challenge*. Routledge.

Barrow, L. (2006). A brief history of inquiry-From Dewey to standards. *Journal of Science Teacher Education, 17*, 265–278.

Becher, T. (1994). The significance of disciplinary differences. *Studies in Higher Education, 19*(2), 151–161. https://doi.org/10.1080/03075079412331382007

Bereiter, C. (1992). Referent-centred and problem-centred knowledge: Elements of an educational epistemology. *Interchange, 23*(4), 337–361.

Berland, L. K., & Steingut, R. (2016). Explaining variation in student efforts towards using math and science knowledge in engineering contexts. *International Journal of Science Education, 38*(18), 2742–2761.

Brodie, K. (2010). *Teaching mathematical reasoning in secondary school classrooms*. Springer.

Bybee, R. W. (1997). *Achieving scientific literacy: From purposes to practices*. Heinemann

Bybee, R. W. (Ed.) (2002). Scientific inquiry, student learning, and the science curriculum. In R. W. Bybee (Ed.). *Learning science and the science of learning* (pp. 25–35). NSTA press.

Bybee, R. W. (2019). Using the BSCS 5E instructional model to introduce STEM disciplines. *Science and Children, 56*(6), 8–12.

Cohen, M., & Hanagan, M. (1991). Work, school, and reform: A comparison of Birmingham, England, and Pittsburgh, USA, 1900–1950. *International Labor and Working-Class History*, *40*, 67–80.

Costa, A., Kallick, B., & Zmuda, A. (2022). *Students: Habits of mind explanation*. The Institute for Habits of Mind. Retrieved from https://www.habitsofmindinstitute.org/wp-content/uploads/2022/02/Student-HOM-Explanation.pdf

Dare, E. A., Ring-Whalen, E. A., & Roehrig, G. H. (2019). Creating a continuum of STEM models: Exploring how K-12 science teachers conceptualize STEM education. *International Journal of Science Education*, *41*(12), 1701–1720.

Delahunty, T., Seery, N., & Lynch, R. (2020). Exploring problem conceptualization and performance in STEM problem solving contexts. *Instructional Science*, *48*(4), 395–425.

Devlin, K. (2000). The four faces of mathematics. In M. J. Burke & F. R. Curcio (Eds.), *Learning mathematics for a new century: 2000 yearbook of the national council of teachers of mathematics* (pp. 16–27). NCTM.

Dewey, J. (1910). Science as subject-matter and as method. *Science*, *31*, 121–127.

Dewey, J. (1938). *Experience and education*. Macmillan.

Duschl, R. A., & Grandy, R. E. (2008). *Teaching scientific inquiry: Recommendations for research and implementation*. BRILL.

English, L. (2023). Ways of thinking in STEM-based problem solving. *ZDM – Mathematics Education*, 1–12. https://doi.org/10.1007/s11858-023-01474-7

Eshach, H. (2008). *Science literacy in primary schools and pre-schools*. Springer.

Friesen, S., & Scott, D. (2013). Inquiry-based learning: A review of the research literature. *Alberta Ministry of Education*, *32*, 1–32.

Gillard, D. (2011). *Education in England: A Brief History*. Retrieved from www.education-uk.org/history

Grant, S. G., Swan, K., & Lee, J. (2017). *Inquiry-based practice in social studies education: Understanding the inquiry design model*. Routledge.

Henningsen, M., & Stein, M. K. (1997). Mathematics tasks and student cognition: Classroom-based factors that support and inhibit high-level mathematical thinking and reasoning. *Journal for Research in Mathematics Education*, *28*(5), 524–549.

Isoda, M., & Katagiri, S. (2012). *Mathematical thinking: How to develop it in the classroom* (Vol. 1). World Scientific.

Just, J., & Siller, H.-S. (2022). The Role of mathematics in STEM secondary classrooms: A systematic literature review. *Education Sciences*, *12*(9). https://doi.org/10.3390/educsci12090629

Kanter, D. E. (2010). Doing the project and learning the content: Designing project-based science curricula for meaningful understanding. *Science Education*, *94*(3), 525–551.

Kelley, T. R., & Knowles, J. G. (2016). A conceptual framework for integrated STEM education. *International Journal of STEM Education*, *3*(1), 11. https://doiorg.ezproxy.lib.purdue.edu/10.1186/s40594-016-0046-z

Kirsh, D. (2009). Problem solving and situated cognition. In P. Robbins & M. Aydede (Eds.), *The Cambridge handbook of situated cognition*. Cambridge University Press.

Korea Foundation for the Advancement of Science and Creativity. (2016). *Introduction to STEAM education*. KOFAC.

Lester, F. Jr., & Kehle, P. E. (2003). From problem solving to modeling: The evolution of thinking about research on complex mathematical activity. In R. Lesh, &

H. M. Doerr (Eds.). *Beyond constructivism: Models and modeling perspectives on mathematical problem solving, learning, and teaching*. Lawrence Erlbaum.

Madden, M. E., Baxter, M., Beauchamp, H., Bouchard, K., Habermas, D., Huff, M., Ladd, B., Pearon, J., & Plague, G. (2013). Rethinking STEM education: An interdisciplinary STEAM curriculum. *Procedia Computer Science*, *20*, 541–546.

Mansilla, V. B., & Gardner, H. (2008). Disciplining the mind. *Educational Leadership*, *65*(5), 14–19.

Mason, J., Burton, L., & Stacey, K. (1985). *Thinking mathematically* (Rev. ed.). Addison-Wesley.

Metiri Group & NCREL. (2003). *EnGauge 21st century skills: Literacy in the digital age*. NCREL.

Moore, T. J., Johnson, C. C., Peters-Burton, E. E., & Guzey, S. S. (2015). The need for a STEM road map. In C. C. Johnson, E. E. Peters-Burton, & T. J. Moore (Eds.) *STEM road map: A framework for integrated STEM education* (pp. 3–12). Routledge.

Nadelson, L. S., & Seifert, A. L. (2017). Integrated STEM defined: Contexts, challenges, and the future. *The Journal of Educational Research*, *110*(3), 221–223.

National Academy of Engineering and National Research Council [NAE & NRC]. (2009). *Engineering in K-12 education: Understanding the status and improving the prospects*. National Academies Press.

National Council of Teachers of Mathematics. (2000). *Principles and standards for school mathematics*. Author.

National Research Council [NRC]. (1996). *National science education standards*. National Academy Press.

National Research Council [NRC]. (2000). *Inquiry and the national science education standards*. National Academy Press.

National Research Council [NRC]. (2011). *Successful K-12 STEM education: Identifying effective approaches in science, technology, engineering, and mathematics*. National Academies Press.

NGSS Lead States. (2013). *Next generation science standards: For states, by states*. The National Academies Press.

Organization for Economic Cooperation and Development. (2005). *The definition and selection of key competencies: Executive summary*. OECD.

Park, W., Wu, J. Y., & Erduran, S. (2020). Investigating the epistemic nature of STEM: Analysis of science curriculum documents from the USA using the family resemblance approach. In J. Anderson & Y. Li (Eds.), *Integrated approaches to STEM education* (pp. 137–155). Springer.

Partnership for 21st Century Skills. (2006). *A state leader's action guide to 21st century skills: A new vision for education*. Partnership for 21st Century Skills.

Poon, C. L., Lee, Y. J., Tan, A. L., & Lim, S. S. (2012). Knowing inquiry as practice and theory: Developing a pedagogical framework with elementary school teachers. *Research in Science Education*, *42*(2), 303–327.

Puntambekar, S., & Kolodner, J. L. (2005). Toward implementing distributed scaffolding: Helping students learn science from design. *Journal of Research in Science Teaching: The Official Journal of the National Association for Research in Science Teaching*, *42*(2), 185–217.

Ring, E. A., Dare, E. A., Crotty, E. A., & Roehrig, G. H. (2017). The evolution of teacher conceptions of STEM education throughout an intensive professional development experience. *Journal of Science Teacher Education*, *28*(5), 444–467.

Rinke, C. R., Gladstone-Brown, W., Kinlaw, C. R., & Cappiello, J. (2016). Characterizing STEM teacher education: Affordances and constraints of explicit STEM preparation for elementary teachers. *School Science and Mathematics*, *116*(6), 300–309.
Ross, V. (2003). The Socratic method: What it is and how to use it in the classroom. *Stanford University Newsletter on Teaching*, *13*(1), 1–4.
Sanders, M. (2009). STEM, STEM Education, STEMmania. *The Technology Teacher*, December/January, *68*(4), 20–26.
Schoenfeld, A. H. (1992). Learning to think mathematically: Problem solving, metacognition, and sense-making in mathematics. In D. Grouws (Ed.), *Handbook for research on mathematics teaching and learning* (pp. 334–370). MacMillan.
Schwab, J. S. (1962). The Teaching of Science as Enquiry. In J. J. Schwab & P. F. Brandwein (Eds.). *The teaching of science*, (pp.1–103). Harvard University Press.
Short, K., Harste, J., & Burke, C. (1996). *Creating classrooms for authors and inquirers* (2nd ed.). Heinemann.
Stohlmann, M., Moore, T. J., & Roehrig, G. H. (2012). Considerations for teaching integrated STEM education. *Journal of Pre-College Engineering Education Research (J-PEER)*, *2*(1), 4.
Tan, A. L., Ng, Y. S., Koh, J. L.-C., Ong, Y. S., & Koh, D. J. Q. (2023b). Applying concepts of plant nutrition in the real-world: Designing vertical farming systems. *Science Activities*, *60*(1), 25–31. https://doi.org/10.1080/00368121.2022.2138249
Tan, A. L., Ong, Y. S., Ng, Y. S., & Tan, J. H. J. (2023a). STEM problem solving: Inquiry, concepts, and reasoning. *Science & Education*, *32*, 381–397.
Tan, A. L., Teo, T. W., Choy, B. H., & Ong, Y. S. (2019). The STEM quartet. *Innovation and Education*, *1*(1), 1–14.
Teo, T. W., & Choy, B. H. (2021). STEM Education in Singapore. In O. S. Tan, E. L. Low, E. G. Tay, & Y. K. Yan (Eds.), *Singapore math and science education innovation* (pp. 43–59). Springer.
Teo, T. W., Tan, A. L., Ong, Y. S., & Choy, B. H. (2021). Centricities of STEM curriculum frameworks: Variations of the STEM Quartet. *STEM Education*, *1*(3), 141–156.
Vasquez, J., Sneider, C., & Comer, M. (2013). *STEM lesson essentials, grades 3–8: Integrating science, technology, engineering, and mathematics*. Heinemann
Wang, S., Ching, Y. H., Yang, D., Swanson, S., Baek, Y., & Chittoori, B. (2020). Developing US elementary students' STEM practices and concepts in an afterschool integrated STEM project. In J. Anderson, & Y. Li (Eds.), *Integrated approaches to STEM education* (pp. 205–226). Springer.
Yeo, J. B. W. (2017). Use of open and guided investigative tasks to empower mathematics learners. In B. Kaur & N. H. Lee (Eds.), *Empowering mathematics learners* (Association of Mathematics Educators 2017 Yearbook) (pp. 219–247). World Scientific.
Yeo, J. B. W., & Choy, B. H. (2023). Fostering disciplinary thinking through mathematical inquiry. *The Mathematician Educator*, *4*(2), 125–147.
Yeo, J. B. W., & Yeap, B. H. (2010). Characterising the cognitive processes in mathematical investigation. *International Journal for Mathematics Teaching and Learning*. Online journal available at http://www.cimt.org.uk/journal
Zimmerman, C. (2000). The development of scientific reasoning skills. *Developmental review*, *20*(1), 99–149.

2 Planning for integrated STEM

Infusing engineering and technology in science and mathematics activities transforms the learning experiences of traditional science and mathematics inquiry from a focus on logical reasoning and explanations to a focus on problem solving. As discussed in Chapter 1, mathematics education has traditionally paid attention to specialising, conjecturing, justifying, and generalising; science education has traditionally valued understanding and explaining natural phenomena, while engineering concentrates on applying scientific knowledge to design and create solutions to real-world problems. How can we then weave the disciplinary knowledge, skills, and practices of engineering and technology with mathematics and science? Building on our discussion on disciplinary features and inquiry in Chapter 1, this chapter delves into instructional frameworks that can be meaningfully applied for planning of integrated STEM learning. To this end, the first section examines how mathematics and science practices can be extended to engineering practices detailed by NGSS, and the second section introduces three variants of the STEM Quartet instructional framework to guide the integration of various practices for integrated STEM inquiry.

2.1 How science and engineering are interwoven in NGSS

Recent educational innovations aimed at integrating different disciplines have emerged to provide a more holistic and practical learning experience, emphasising both conceptual understanding and application in our everyday lives. The global shift towards this interdisciplinary integration has been implemented in educational curricula and national standards around the world. A prime example of this transformation is the Next Generation Science Standards (NGSS).

How can engineering practices be incorporated in science inquiry? In response to this question, NGSS has established the concepts of 'science and engineering practices' by integrating engineering practices into science practices (NGSS Lead States, 2013). The conceptualisation of 'science and engineering practices' in NGSS is fundamentally connected to the basic idea of STEM education, involving the merging of ideas, skills, and practices of different disciplines such as science, engineering, and technology. In this vein, let's take a closer look at science and engineering practices in NGSS.

There are a total of eight science and engineering practices in NGSS: (1) asking questions and defining problems; (2) developing and using models; (3) planning and carrying out investigations; (4) analysing and interpreting data; (5) using mathematics and computational thinking; (6) constructing explanations and designing solutions; (7) engaging in argument from evidence; and (8) obtaining, evaluating, and communicating information. As 'science and engineering practices' indicates, the eight practices encompass characteristics of both scientific and engineering practices.

Fundamentally, the eight practices can be interpreted and implemented with science-focused and engineering-focused aspects. This is because both scientific and engineering practices involve these eight practices, albeit from slightly different perspectives. Taking the first practice (asking questions and defining problems) as an example, scientific inquiry involves the formulation of a question that can be answered through investigations of the natural world and how it works, while engineering design involves the formulation of a problem that can be solved through design. Table 2.1 summarises the different emphases of the eight practices in scientific practice and engineering practice (Cunningham & Carlsen, 2014).

The differences and similarities between science and engineering practices highlighted in Table 2.1 can be attributed to different disciplinary values or goals. For instance, science practices use the scientific method to construct models for explaining and predicting natural phenomena, while engineering practices use the scientific method to find patterns or verify solutions in solving problems. However, despite having different perspectives, we can still see that science and engineering practices share several core similarities from a practice perspective. For instance, both practices share similar approaches to problems by relying on evidence, employing mathematical and computational thinking, and communicating results. Both fields also engage in iterative processes of testing and apply evidence for improvement.

The attributes of science and engineering practices can be applied in various ways when planning integrated STEM learning experiences. For example, Lilly et al. (2022) categorised NGSS-aligned lessons into three types—science-focused, engineering-focused, and computational thinking-focused lessons—based on the practices used in the lessons. They reported that lesson flows and ways of pedagogical support differed depending on which practices formed the focus of a lesson. For instance,

Table 2.1 Differences in the practices of science and engineering (Cunningham & Carlsen, 2014)

Practices in NGSS	Relative emphasis in science	Relative emphasis in engineering
(1) Asking questions and defining problems	Goal is theoretical/conceptual progress	Goal is a useful, novel technology
(2) Developing and using models	To enable explanation and prediction	To facilitate analysis and evaluation
(3) Planning and carrying out investigations	Hypothesis-testing, which may be sequential	For evaluation of prototypes or ideas, usually iterative
(4) Analysing and interpreting data	Attention given to measurable aspects of both the discovered and natural world	Attention given to more diverse criteria including scientific (e.g., material properties) and others (e.g., cost, risk of failure)
(5) Using mathematics and computational thinking	Testing conceptual models with real data	Designing concrete things, using both real and simulated data
(6) Constructing explanations and designing solutions	Objective is a single 'best explanation'	Objective is a preferred design, selected among alternatives, with explicit consideration of tradeoffs
(7) Engaging in argument from evidence	Goal is to persuade scientific peers	Goal is to satisfy a client
(8) Obtaining, evaluating, and communicating information	Free exchange of information is an important norm	Products are often legally proprietary and information guarded

during science-focused lessons, teachers provided considerable support to students in the practices of constructing explanations and engaging in argument from evidence. However, in engineering-focused lessons, the pedagogical emphasis tended to shift towards assisting students extensively in defining problems.

Similar to ideas presented by Lilly et al. (2022), integrated STEM learning can also be designed in a myriad of ways based on the disciplinary practices incorporated and their role in supporting STEM problem-solving processes. The distinguishing characteristics of STEM practices are that several science and engineering practices can be selected and incorporated in a holistic manner to form the STEM problem-solving process. The following section will describe three different instructional models, each based on a different STEM learning focus and matched to the relevant science and engineering practices.

2.2 The STEM Quartet instructional framework of integrated STEM learning

Problem solving is a process that is sensitive to the nature of a problem and thus requires different practices and skills to be meaningfully engaged in the process. The challenge with integrated STEM problem solving lies in deciding which science and engineering practices can be chosen in a systematic manner when encountering a problem. STEM problem-solving practices are not simply a collection of science and engineering practices stacked on top of one another. Rather, it refers to the practices that are interwoven into the whole process of problem solving and the meaningful connections between one another. Teachers may feel disoriented and overwhelmed when designing integrated STEM lessons because there are diverse ways of integrating the different practices of STEM learning (refer to Figures 1.4–1.6 in Chapter 1). How can one decide on the goals and directions of STEM lessons and prioritise the related practices?

There are many ways to design and enact integrated STEM inquiry experiences to engage students' lived experiences within the context of real-world problems. The choice of lesson design should align with the lesson intention(s). Teo et al. (2021) proposed three centricities or starting points for planning integrated STEM inquiry—(1) problem-centric, (2) solution-centric, and (3) user-centric.

In problem-centric STEM inquiry, students start with a persistent, extended, and complex problem and are tasked with generating and designing solutions to the problem (Tan et al., 2019). The problem-solving process involves making sense of the context of the problem and unpacking the components of the problem. This is followed by defining the problem space in which they will work. Typically, these complex, persistent, and extended problems are authentic real-world problems with multiple plausible solutions. Hence, different student groups may work on different aspects of the problem.

When engaging in solution-centric STEM inquiry, students start with existing solutions to a problem and are required to examine the affordances, advantages, and disadvantages of the solutions. They subsequently engage in the improvement, design, and enhancement of a solution, while considering current limitations. They test their new designs and continue to refine their solutions using the evidence collected. As a solution-centric learning experience is characterised by starting with a solution rather than the problem, it carries features of design-based learning.

User-centric STEM inquiry focuses on understanding the user and creating solutions that address specific goals and needs of specific users within their context of use (Teo et al., 2021). In user-centric STEM inquiry, the users' evaluation of the proposed solutions serves as feedback for students to improve their solutions.

The different centricities of STEM inquiry result in varied learning experiences. Tan et al. (2023) compared the types of questions asked, ways of reasoning, and creativity of students when two groups of students were presented with similar STEM activities with one group starting from a problem and the other starting from existing solutions. They found that in both problem-centric and solution-centric classrooms, ontic questions (questions seeking to understand the observable entities that make up a concept) were most common, while epistemic questions (questions seeking justification and ideas or information to help understand how we know what we know) were least used.

For reasoning, practical, or commonsense reasoning (reasoning based on practical success criteria) was more frequently used compared to scientific reasoning (reasoning based on a conceptual value system) in STEM problem solving. As context and content are both important in integrated STEM inquiry, understanding the context without understanding the scientific principles of how things work does not contribute to meaningful learning.

For creativity, creativity test scores showed that students in problem-centric classes showed higher problem-solving abilities, scientific imagination, and creative product design compared with the solution-centric STEM inquiry group. However, the problem-centric group showed weaker performance in their ability to improve a technical product and demonstrated lower creative experimental ability and overall creativity gains.

The observations made by Tan et al. (2023) suggest that presenting students with different starting contexts has an impact on their STEM inquiry experiences. The choice of lesson focus should depend on factors such as the learning intention(s), availability of resources, and the level of readiness of students. For instance, if students are experiencing integrated STEM inquiry for the first time or if they are from lower grade levels, it might be better to start them with a solution-centric STEM inquiry. This is because solution-centric activities tend to be more convergent compared to problem-centric activities, which typically require students to decipher the 'noise' from the evidence of the problem, with the latter requiring higher cognitive engagement than the former. In the next section, we discuss some variations that can be considered in planning integrated STEM inquiry. The focus and features of the three variants of the STEM Quartet instructional framework are described next.

2.2.1 Problem-centric variant

The problem-centric variant starts with a problem and focuses on problem-solving processes: identifying and defining a problem and generating solutions. This model describes how students solve complex, extended,

Planning for integrated STEM 33

and persistent problems in six stages of the STEM inquiry process as shown in Figure 2.1.

In this variant, students are first asked to establish, understand, and identify their STEM problems within a specific everyday context. Depending on the requirements of the problem presented, any of the four disciplines can be the lead discipline to start the formulation of possible solutions. During the process of identifying and understanding the problem, students make conjectures of the relationships between different variables associated with the problem. They make decisions regarding the main cause of the problem and the incidental factors causing the problem. Identifying and understanding the problem are important first steps for students. Identification of the 'wrong' problem would likely result in added problems. Tan et al. (2023) observed that students in problem-centric lessons tended to spend more time in discussion, debates, and justification to determine the problem compared to students in solution-centric classes. Once the problem is identified, students start working on designing, refining, and testing their solution.

In the design of the solution, students apply specific disciplinary knowledge and skills. In the process of generating solutions, students engage with brainstorming and generative discussions, and design, evaluate, test, review, and improve possible solutions. As real-world problems are complex and unlikely to be solved with the knowledge and skills of a single discipline, the problem-solving process affords students to select from, apply, and connect the knowledge, practices, and social norms between

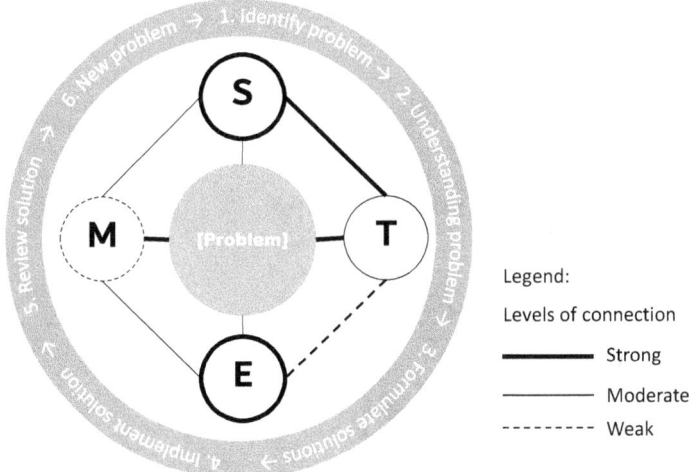

Figure 2.1 The problem-centric STEM learning process (Tan et al., 2019).

34 *Planning for integrated STEM*

the four STEM disciplines as they work on plausible solutions. In Figure 2.1, we can see the connections between the four disciplines with the problem seated at the centre. Depending on the problem and its components, the nature of the connections between the disciplines may differ.

2.2.2 Solution-centric variant

A solution-centric lesson typically starts with a critical and realistic issue or existing ways of doing things to focus students' attention on the shortfalls or potential areas for improvement of established or existing solutions. In this variant, the focus is on finding the most appropriate and improved solution through the iterative processes of evaluating and refining or enhancing current solutions. By evaluating existing solutions, students learn the knowledge and skills of different disciplines that have been applied to make existing solutions work. Figure 2.2 describes how students engage in this solution-centric STEM learning.

When presented with existing solutions, students evaluate and test the gaps in existing solutions to generate a new, improved solution. Specific to this process, students can engage more meaningfully in the 'looking back'

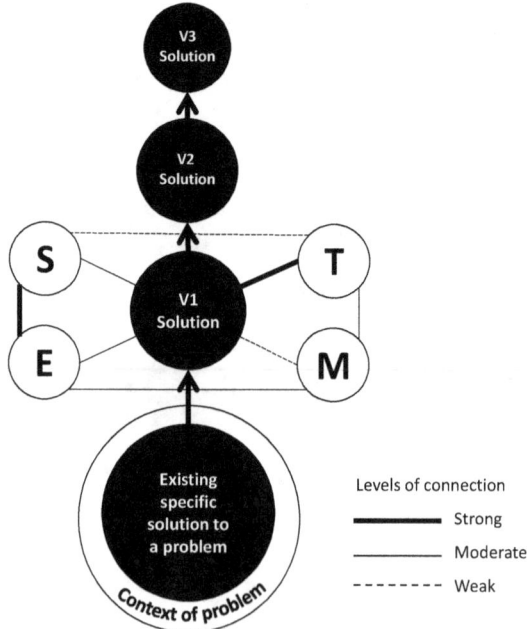

Figure 2.2 The solution-centric STEM learning process (Teo et al., 2021).

(Teo et al., 2021, p.144) phase of design improvement. As illustrated in Figure 2.2, the practices of the four disciplines are interconnected to one another instead of the problem.

2.2.3 User-centric variant

The last variant of the STEM Quartet framework is user-focused. The user-centricity focuses first on understanding and meeting the needs and preferences of the users. This framework guides students to empathise with the problem faced by the users and interpret it from the user's point of view. Students are encouraged to analyse the problem in the context of the users by identifying the users' pain points, needs, and desires. In this vein, the user-centric framework helps students see the interaction between the human and technology, and the resulting influence to consider the 'humanistic' side of real-life problems. Figure 2.3 summarises the learning processes of a user-centric variant.

In Figure 2.3, the practices of science, engineering, technology, and mathematics surround the user, meaning that the four discipline practices are applied to understand and address the user's specific contexts and needs. Once the user needs are well understood, the focus shifts to finding a solution that effectively meets those needs. This model is often used in the fields of design and user experience where the focus is on creating products and services that are user-friendly and meet the needs of the users.

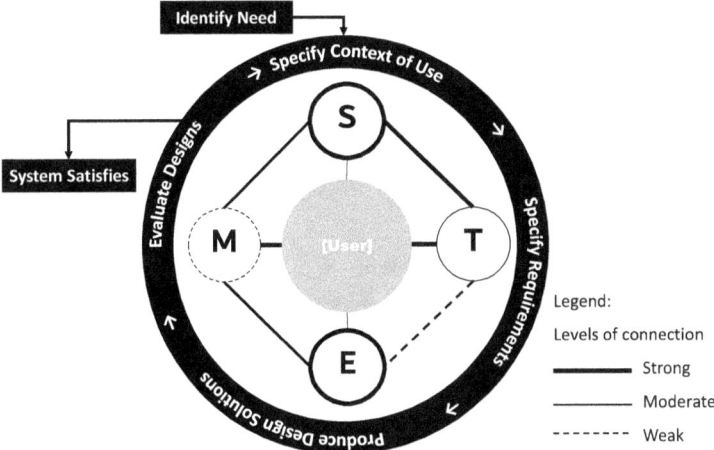

Figure 2.3 The user-centric STEM learning process (Teo et al., 2021).

2.2.4 Choice of variant

While all three variants of the STEM Quartet framework have their merits, they differ in terms of where the initial focus lies and how the problem-solving process unfolds. The problem-centric variant places primary importance on understanding the problem, the solution-centric variant focuses on improving an existing solution, and the user-centric variant prioritises understanding and meeting user needs. Table 2.2 summarises the focus and knowledge types emphasised in each variant.

The choice of variant will depend on the nature of the STEM problem, the available resources, students' level of readiness, and the desired learning outcome(s). It is crucial to recognise that the three STEM instructional models are not mutually exclusive but rather provide different perspectives to enrich and diversify problem-solving processes as STEM practices. In this context, flexibility is a key attribute for teachers, enabling them to navigate these options effectively, make informed decisions, and tailor their approaches to the unique demands and objectives of each classroom situation. In a study by Tan et al. (2022), where the types of questions asked by students, the levels of argumentation, and the creativity between problem-centric and solution-centric lessons were compared, the authors arrived at three key conclusions. First, in both problem- and solution-centric lessons, ontic questions (questions seeking to understand the observable entities that make up a concept) were most asked, while epistemic questions (questions seeking justifications and ideas or information to help understand how we know what we know) were least asked.

Table 2.2 Comparison of the problem-, solution-, and user-centric variants (Teo et al., 2021)

Variant	Focus	Knowledge types prioritised in the 21CC framework
Problem-centric	Complex, extended, and persistent problem	Meta Knowledge: Students may think of different ways to solve the problem creatively and collaboratively
Solution-centric	An existing solution to (part) of a complex, extended, and persistent problem	Foundational Knowledge: The solution may be well defined and core content knowledge and cross-disciplinary knowledge are preidentified (e.g., use of technology as a requirement).
User-centric	Existing and potential users of the STEM solution output(s)	Humanistic Knowledge: Development of empathy in designers can be an outcome of the process.

Second, practical reasoning was more frequently used than scientific reasoning to support argumentation during STEM problem solving in both types of lessons. Lastly, the creative abilities of the students were measured using a modified version of the scientific creativity test for secondary school students by Hu and Adey (2002). In terms of creativity, differences were observed between problem-centric and solution-centric lessons. Students in problem-centric classes demonstrated stronger problem-solving abilities but performed weaker in their abilities to improve a technical product and experiment creatively and in their overall creativity gains compared with students in solution-centric classes.

2.3 Mapping STEM practices to disciplinary expertise

The three variants of the STEM Quartet instructional framework described earlier help teachers decide if they want to focus on problem solving, design improvement, or enhancing user experiences. The next step involves deciding on the lead discipline for a lesson. Teachers do not have all the expertise required for all four disciplines of STEM. As such, one way to make integrated STEM lessons possible is to have a lead discipline. A lead discipline serves as the starting point for a STEM activity and directs how the other disciplines are connected. For instance, one can refer to Figure 1.5, where science can be the lead discipline to introduce the problem of lack of clean water, develop students' ideas on different water treatment systems, and direct students to learn about different types of pollutants and their effects on human health. The science teacher can collaborate with the mathematics teacher to design a lesson on how data related to different designs of water treatment systems and their corresponding pollutants can be measured and represented. The technology teacher can contribute by teaching students how to use sensors to detect changes in water quality such as increased dissolved carbon dioxide concentration. In another instance, if the mathematics teacher is to lead the activity, he/she could start the activity by presenting students with a dataset on diarrhoea cases in different communities and living conditions across different age groups. The teacher could guide students to look for patterns and create a mathematical model showing the relationship between the various factors and probability of getting diarrhoea. Upon the development of this idea, the mathematics teacher can approach the science teacher to invite him/her to engage students in discussions on different living conditions and the potential exposure to water-borne pathogens. Similarly, if a technology teacher leads the STEM lesson, he/she can start by introducing students to sensors such as dissolved oxygen sensors, dissolved carbon dioxide sensors, and pH sensors. Students can learn how to connect these sensors to collect data on changes in the physical environment. The mathematics teacher can then introduce students to different types of graphical representations suitable for the data collected. Additionally, the science teachers could follow up by having the students inquire about the different conditions that could lead to changes in oxygen, carbon dioxide, and pH levels of the water. Hence, it is evident that even though the problem presented is the same, the resultant learning experiences are different when different disciplines serve as the lead discipline since the epistemic practices and goals of each discipline are differently emphasised. Table 2.3 maps the variations of each discipline-led practice against the three variants of the STEM quartet framework.

2.4 Mapping specific STEM practices

In Table 2.3, we observe that the science-led solution-centric STEM activities prioritise the activity of researching existing scientific concepts related to a proposed solution and conducting experiments to test the feasibility of solutions. How could a science-led solution-centric lesson be enacted? Do all science-led solution-centric lessons look the same? The obvious answer is no.

The classroom practices and strategies chosen for each lesson are different depending on the nature of the STEM problem, the intended learning outcome(s), the level of readiness of the students, and the depth of content knowledge that the students have. Table 2.4 shows a matrix mapping the STEM practices against possible variations in a solution-centric STEM lesson. Specifically, the practices in the shaded boxes are examples of practices contextualised within the solution-centric framework. For example, the practice of 'asking questions' can encompass asking scientifically oriented, procedural, or practical questions. In the solution-centric model, asking practical questions often holds more relevance. The choice of questions to ask students in a solution-centric activity is also dependent on student readiness. For instance, if students are encountering integrated STEM inquiry for the first time, teachers could present them with practical questions that students are more familiar with compared with scientifically oriented questions.

Similarly, for the practice of data analysis, it can involve discerning cause and effect, describing relationships, and identifying trends. If students have strong mathematical abilities, the activity can be designed such that the students are required to choose an analytic method that best describes the relationship between the different variables. Besides students' abilities, the choice of analysis is also dependent on the requirement(s) of the task set by teacher(s). For instance, if students are given particles of different sizes and asked to establish which particle would be most water retentive, the data analytic task would require students to analyse and identify trends. As such, teachers decide on the focus of each practice after considering the nature of the problem, the level of readiness of the

Table 2.3 Variations of science-, technology-, engineering-, and mathematics-led practices mapped to different centricities

STEM practices	Problem-centric STEM activities	Solution-centric STEM activities	User-centric STEM activities
Science-led practices	- Explaining scientific aspects related to a problem. - Drawing conclusions as a solution based on the experiment conducted.	- Researching existing scientific concepts related to a proposed solution. - Conducting experiments to test the feasibility of the solution.	- Researching the user's interaction with possible solution. - Conducting experiments to validate the solutions.
Engineering-led practices	- Designing, testing, and refining prototypes. - Implementing the most efficient solution.	- Optimising the design and functionality of a proposed solution. - Applying engineering principles to improve the quality of solutions.	- Identifying user needs. - Designing and iterating prototypes based on user feedback. - Optimising the solution for usability and accessibility.
Technology-led practices	- Using hardware or software for prototyping and testing. - Using technology as a tool for communication and collaboration.	- Developing prototypes. - Using technology for data analysis to improve solutions.	- Using technology to collect more accurate information from users. - Using technology to empower users. - Employing software for user interface design.
Mathematics-led practices	- Using models to represent and predict the possible outcomes of problems. - Applying trends and patterns to create solutions.	- Developing improved models with greater predictive validity with additional information.	- Using data analytics to analyse user behaviour.

Planning for integrated STEM

Table 2.4 An example of mapping STEM practices to a solution-centric lesson

Practices	Variations and concretisation of STEM practices		
Asking questions	□ Scientifically oriented questions	□ Procedural questions	■ Practical questions
Data analysis	□ Analysis to identify cause and effect	■ Analysis to describe relationships	□ Analysis to show trends
Creativity	■ Cognitive creativity	□ Process creativity	□ Product creativity
Critical thinking	■ Rationale-based inference	□ Model-based inference	□ System-based inference
Collaborative problem solving	□ Interest sharing	■ Role sharing	□ Collaborative monitoring
Communication	□ Description of opinion	■ Expression of opinion	□ Product criticism

students, resource availability, and the intended learning experiences. Consequently, given the same problem, the lesson enactment by different teachers could be different as each teacher could choose a different combination of variation for each practice. Table 2.4 can serve as a guide to teachers as they plan different integrated STEM learning experiences for varying contexts and students.

In summary, the STEM practices of asking questions, data analysis, creativity, critical thinking, collaborative problem solving, and communication are deeply rooted in science, engineering, technology, and mathematics, and form the core of STEM lessons. When implemented in the classroom, these core STEM practices can vary depending on factors such as the nature of the STEM problem encountered, students' level of readiness, resource availability, and the intended learning experiences. Another factor that could result in variations in STEM lesson enactment is whether a lesson is planned to have a science, engineering, mathematics, or technological focus. The variations of different STEM practices that can be mapped against the three variants of the STEM Quartet instructional framework suggest that teachers must remain flexible and adaptable when planning and enacting integrated STEM learning experiences. They must be sensitive to the learning contexts, the problem requirements, and the knowledge level of students when planning their lessons. In Chapters 6–10, we present practices contextualised in actual

STEM lessons to illustrate how STEM practices can be enacted differently in different classroom contexts. Identifying specific STEM practices in the planning process also enables teachers to use them as evaluation criteria for assessing STEM learning. How these STEM practices can be used as evaluation indicators will also be discussed in detail in the next chapter.

Questions for Reflection

1 Which variant of the integrated STEM Quartet instructional framework appeals to you? Why?
2 Describe how the proposed integrated STEM practices are similar or different from the practices of science?
3 Why is it important to allow for variations in STEM practices?

References

Cunningham, C. M., & Carlsen, W. S. (2014). Teaching engineering practices. *Journal of Science Teacher Education*, *25*(2), 197–210.

Hu, W., & Adey, P. (2002). A scientific creativity test for secondary school students. *International Journal of Science Education*, *24*(4), 389–403. https://doi.org/10.1080/09500690110098912

Lilly, S., McAlister, A. M., Fick, S. J., Chiu, J. L., & McElhaney, K. W. (2022). Elementary teachers' verbal supports of science and engineering practices in an NGSS-aligned science, engineering, and computational thinking unit. *Journal of Research in Science Teaching*, *59*(6), 1035–1064.

NGSS Lead States. (2013). *Next generation science standards: For states, by states*. The National Academies Press.

Tan, A.-L., Ong, Y. S., & Ng, Y. S. (2022). *Comparing students' questioning, argumentation, and creative thinking in problem-centric and solution-centric STEM activities*. Unpublished technical report. National Institute of Education: Office of Education Research.

Tan, A. L., Ong, Y. S., Ng, Y. S., & Tan, J. H. J. (2023). STEM problem solving: Inquiry, concepts, and reasoning. *Science & Education*, *32*, 381–397.

Tan, A. L., Teo, T. W., Choy, B. H., & Ong, Y. S. (2019). The STEM quartet. *Innovation and Education*, *1*(1), 1–14.

Teo, T. W., Tan, A. L., Ong, Y. S., & Choy, B. H. (2021). Centricities of STEM curriculum frameworks: Variations of the STEM quartet. *STEM Education*, *1*(3), 141–156.

3 Disciplinarity in integrated STEM inquiry

In Chapter 1, we learnt about the diverse conceptions of integrated STEM held by teachers and appreciated the fact that integrated STEM inquiry shares similar epistemic practices (such as formulation of questions, logical deduction, and testing of ideas) with science and mathematical inquiry. However, integrated STEM inquiry differs from science and mathematical inquiry with regard to the object of inquiry—real-world problems that require the application of practical and disciplinary knowledge to solve them. While we acknowledge and celebrate the variations in teachers' conception of integrated STEM inquiry, highlighting the similarities in ideas and practices is important for the community to develop common understanding of integrated STEM inquiry. In Chapter 2, we described the planning of integrated STEM inquiry using the STEM Quartet instructional framework. Whether we choose a problem, solution, or the user as the starting point, the learning experiences planned take into consideration how specific disciplinary knowledge is connected to one another. In this chapter, we will delve into the disciplinarity of integrated STEM inquiry to distil the defining characteristics of integrated STEM inquiry and its enactment. Understanding the disciplinarity associated with integrated STEM inquiry will enable teachers to reflect on their professional actions and decisions. The questions that we seek to answer are 'What are the defining disciplinary features of integrated STEM inquiry?' and 'What are the different ways to reflect on enactment of integrated STEM inquiry?'

3.1 Discovering disciplinarity

Disciplines are defined by unique knowledge and practices which are marked by discipline boundaries. The study of science and mathematics has a long history as disciplines with unique epistemic, social, and cognitive structures. These robust disciplinary practices suggest that disciplinary boundaries are strong and compact and hence not porous to ideas from other disciplines. These compact disciplinary boundaries pose problems for integration.

Ford and Forman (2006) discussed the three disciplinary features of science classrooms as the (1) material aspects of science, (2) social aspects of science, and (3) the interplay of roles between a Constructor and Critiquer of claims. By material aspects of science, they refer to the use of tools, equipment, and manipulatives for testing, collecting data, and observation. The social aspects of science describe how communities of learners engage in group collaborations, discussions, and debates. Finally, creating opportunities for students to present their ideas and claims in a public sphere for peer critique is also an important consideration. In integrated STEM learning, the requirement for engagement with real-world problems or issues is also an important disciplinary feature. Adapting Ford and Forman's (2006) disciplinary ideas, we propose consideration of the following aspects for integrated STEM inquiry:

1. Connection to real-world problems;
2. Opportunities for presentation and critique of ideas and claims (inter-play of role as Constructor and Critiquer of claims);
3. Space for design, making, and investigation (material aspect); and
4. Modelling communities of STEM problem solvers (social aspect).

Aside from the four aforementioned conditions, teachers should also be aware of the differences between monodisciplinary and integrated STEM inquiry pedagogical practices when enacting an integrated STEM inquiry lesson. Being aware of the differences and similarities in pedagogical practices of monodisciplinary and integrated forms of STEM inquiry enables teachers to better keep their practices in check. Pedagogical practices refer to things that teachers do to engage and connect with their students in a positive and productive manner. Ong et al. (2023) proposed a framework comparing the conceptual/procedural, epistemic, and social aspects of monodisciplinary and integrated STEM lesson enactment. Table 3.1 maps the different pedagogical practices of monodisciplinary lessons to those of integrated STEM inquiry. It is noted that in integrated STEM, there are some aspects of pedagogical practices that we are still seeking to illuminate. For instance, in introducing the real-world context and problem in integrated STEM, the epistemic aspect of that practice is still vague and likely to require more discussions and debates for agreement. Where practices are still contentious, we blank out the specific aspect.

Mapping the four proposed conditions to their associated pedagogical practices, it is evident that the connection to real-world problems is aligned with the practice of introducing context and problems, while opportunities for presentation and critique are fulfilled by creating space for students to clarify solution requirements and request for justification(s).

Table 3.1 A framework for integration of STEM pedagogical practices for lesson enactment for teachers (Ong et al., 2023)

Pedagogical practices	Conceptual/procedural aspects	Epistemic aspects	Social aspects
Activating prior knowledge [Monodisciplinary]	Recalling/Introducing scientific or mathematical conceptual knowledge through investigations or lecturing.	Make references to empirical evidence for scientific claims or engage in conjecturing, justifying, or generalising in mathematics.	Bringing personal experiences with scientific or mathematical phenomena into the classroom. Highlight the historical development of scientific or mathematical ideas.
Introducing the real-world context and problem [Integrated STEM]	Recalling/Introducing disciplinary-based conceptual knowledge needed to solve the STEM problem.		Introducing the sociocultural context in which the problem is situated. Highlighting relevant social aspects of the problem, e.g., user/client needs.
Providing clear explanation [Monodisciplinary]	Crafting claims that are supported by empirical evidence and use scientific reasoning to connect the claims to the evidence. For mathematics, use of conjectures and justifying developing conceptual knowledge.	Understand the roles of observations, measurements, and logic to increase our understanding of both the natural and man-made world around us.	Agreement and accountability to a community of scientists or mathematics is required for acceptance of proposed explanations.
Clarifying the solution requirements and requesting for justification(s) [Integrated STEM]		Clarifying success criteria and constraints that the solution needs to fulfil, some of which should be relevant to the applicable STEM discipline. Requesting justification(s) from students to make explicit their decision.	Ensuring that success criteria and constraints should be relevant to the intended user/client. Making group-based STEM problem-solving decisions public for critique and to persuade others of the soundness of the solution.

Disciplinarity in integrated STEM inquiry

Practice	Column 1	Column 2	Column 3
Pacing and maintaining momentum [Monodisciplinary]	Learning progression of scientific or mathematical ideas as concepts and ideas build on one another to develop understanding.	Existing ideas come from earlier ideas and are improved and changed based on fresh evidence.	Engagement in argumentation to justify explanations based on available evidence. Engagement in justifying conjectures and generalisation.
Pacing and maintaining momentum of an iterative problem-solving cycle [Integrated STEM]	Planning student learning of needful disciplinary-based conceptual knowledge/procedural skills at different stages in the problem-solving cycle.	Revisiting stages in the iterative cycle, a necessary process for students to improve their ideas and solve the problem.	Being responsive to group dynamics that may mitigate the pace of the problem-solving cycle.
Concluding the lesson [Monodisciplinary]	Usually involves the presentation of concepts or proofs that have been legitimatised by community of scientists or mathematicians. Students usually demonstrate understanding of concepts through completion of worksheets.	Presentation knowledge as statements of generalisation or mathematical rules for formulae that have been accepted.	Students and teachers clarify the scope and requirements of specific scientific or mathematical concepts.
Reflecting on the STEM learning experience [Integrated STEM]	Engaging students in reflection to make salient the conceptual/procedural, epistemic, and social aspects of their STEM learning experience, to prepare them for future learning.		

Notes: Pedagogical Practices for Monodisciplinary Lesson Enactment.
Pedagogical Practices for Integrated STEM Lesson Enactment.

Having students work in teams to clarify solutions also allows students to experience the social aspect of working in problem-solving communities. The material aspect of STEM inquiry is fulfilled by students working on iterative problem-solving cycles. In the next section, we elaborate on each of the four conditions highlighted.

3.1.1 Condition 1: Connection to real-world problems

The learning opportunities for STEM inquiry are made more relevant when they are presented to students in the form of real-world problems (Sias et al., 2017). By anchoring teaching and learning in real-world problems, students can make connections and appreciate the relevance between the disciplinary content learnt in school and real-life application in their lives. Engagement with real-world problems can often mirror students' lived experiences and present opportunities for them to acquire scientific understanding through solving feasible, worthwhile, contextualised, meaningful, ethical, and sustainable problems (Krajcik, 2015). Students build new knowledge and develop a deeper understanding of the world around them based on what they already know and believe (NRC, 2000). This forms the foundation for initiating change in students' ideas and actions for sustainable development since students have their personal conceptions of natural phenomena that can influence their learning.

Real-world problems, however, are often non-routine. To become an expert problem solver, one needs to be able to identify the right problems and devise plausible solutions. Proposed solutions are both an artefact of conceptually linked information and the result of metacognitive efforts of reflection on the choice and effectiveness of specific problem-solving strategies (Levy & Murnane, 2004). The complexities of problems and the resultant uncertainty of solutions demand that learners consider the context of the problem, the needs of the stakeholders, and the limits and constraints of the proposed solutions. The problem-solving process hence affords learners opportunities to handle ambiguity. To generate new and innovative solutions, creativity is also required to integrate seemingly unrelated information and foresee possibilities others may miss.

Real-world problems can be presented to students in different forms. At the most challenging and complex level are 'wicked problems' (Rittel & Webber, 1973; Buchanan, 1992). Wicked problems are multi-dimensional and complex, and oftentimes do not have a single solution. Similar in characteristics to wicked problems are ill-structured problems. As described by Jonassen (1997), ill-structured problems are phenomena that 'possess multiple solutions, solution paths, fewer parameters which are less manipulable, and contain uncertainty about which concepts, rules, and principles are necessary for the solution or how they are organised and which solution is best' (Jonassen, 1997, p. 65). Jonassen also contrasted ill-structured problems with well-structured problems, which are constrained problems with well-defined parameters and usually fewer alternative solutions.

The 17 United Nations Sustainable Development Goals (SDG) identified in 2015 are examples of wicked or ill-structured problems involving eradication of poverty, reducing hunger, ensuring good health and well-being, affordable and clean energy, industry, innovation and infrastructure, and climate action (United Nations, 2015). These challenges can be introduced to help students relate to the global issues of sustainability and develop their abilities to generate solutions. Working with wicked or ill-structured problems requires students to ask questions and apply knowledge and skills across different domains (Wade et al., 2020). The complexities and demands of wicked or ill-structured problems require guidance for students to break the problems into their component parts and work with bite-size portions of manageable cognitive loads.

Aside from real-world wicked or ill-structured problems, Bereiter (1992) proposed another classification of problems—(1) problems of explanation and (2) transient or practical problems. Problems of explanation are steep in inquiry and require domain-specific knowledge to solve. Problems of explanation serve as organising points for knowledge, consisting of conceptual representations, typically a text base and a situation in which the text is relevant. Transient or practical problems resemble problems that students face in their daily lives. Transient problems disappear once a solution is available; both the problem and solution are not likely to be remembered after that. Practical problems are issues that can be solved with or without basic level knowledge, as a result the solutions often lack precise definition. Described by Bereiter (1992), an example of a practical problem is when a door becomes stuck. Practical solutions would include kicking the door or pushing it harder. These solutions lack references to established and reproducible scientific or mathematical principles and hence may be limited in promoting inquiry and improvement.

The authenticity of the problems presented can also influence the meaningfulness of the problems. As it is not always possible for students to engage in authentic STEM inquiry in the real world to test and apply their ideas, it is useful to consider how the problem presented can be made as authentic as possible. Roach et al. (2018) proposed four dimensions of authenticity that can be applied to understand the authenticity of a STEM problem—(1) context authenticity, which describes ideas that come from real-world problems motivated by students' interactions with their environment; (2) task authenticity, which focuses on processes of inquiry that resemble real-world scientific inquiry activities; (3) impact authenticity, which pays attention to the impact of students' work in the real-world environment; and (4) personal or value authenticity, which measures the extent to which the problem is related to the students' lives and sparks curiosity to motivate them to seek answers. Given the vast variety of problems that can be presented to students, it is important to consider the grade levels, maturity, contexts, and domain knowledge that students possess when designing STEM inquiry experiences. Students' interests and experiences can serve as valuable guides in designing meaningful and authentic problems. Some questions that we can consider when planning and enacting our integrated STEM lessons include: (1) How does the problem relate to the real-world context? (2) Which authenticity dimension is the problem activity designed to fulfil? (3) What strategies can I use to introduce the problem and the context? and (4) How can we meaningfully engage students' prior and current experiences to work on these problems?

Students learn by building on the knowledge and experiences that they already have. Successful STEM inquiry and problem-solving identify relevant STEM background knowledge and preconceptions that students possess and use them as starting points to build greater and deeper understandings. While students' preconceptions and initial knowledge may be limited and inconsistent with science (Driver et al., 1994), uncovering their naïve conceptions through engagement with the STEM inquiry processes offers opportunities for these ideas to be re-negotiated. These preconceptions and everyday experiences can serve as scaffolds for the learning of domain understandings as well as competencies. Similarly, acknowledgement of students' preconceptions and experiences also offers insights into possible barriers in students' learning and acquisition of new concepts (Sinatra & Pintrich, 2002).

To sustain students' interest and experiences, lessons can incorporate opportunities to (1) engage students in the practices of various disciplines, (2) create spaces for students to be involved in out-of-school STEM activities, and (3) incorporate elements of career exploration and readiness within the STEM learning experiences, to foster their aspirations within realistic and authentic contexts. It is important to sustain students' interest by connecting real-world problems to familiar experiences as integrating STEM inquiry and problem solving is a novel approach that requires teachers and students to share authority and autonomy, concepts which can be challenging for young or low-progress learners (Isik-Ercan, 2020).

Let us consider as an example the real-world problem of designing portable water filtration systems in areas where clean water is not readily available. To evoke students' interest in the problem, students can be tasked with taking photos of water filtration devices that they can find within their homes or school. Students can examine these photos and raise questions about the relative sizes of the water filtration devices in comparison to the number of users, before considering the similarities and differences in the component parts (i.e., the filters and their efficiency) of filters found in different locations. Students can also inquire how the efficiency of filters is evaluated and draw connections between the design and role of the filters. In order to engage students' interest and knowledge, we can consider asking ourselves two questions when planning and enacting STEM lessons: (1) How is the problem related to the real-world context? (2) Which authenticity dimension is the problem activity designed to fulfil?

3.1.2 Condition 2: Opportunities for presentation and critique of ideas and claims (interplay of roles as constructor and critiquer of claims)

To enable students to experience the role of constructor and critiquer of claims during problem-solving and inquiry processes, both the teachers and students need sound domain-specific knowledge and skills (Priemer et al., 2020). Research on how experts function revealed that experts demonstrate deep foundational understanding of factual knowledge within a field and are able to organise and connect these knowledges to facilitate retrieval and application (Donovan et al., 1999). Further, since the process of problem solving requires students to integrate disciplinary content knowledge that is usually coupled with procedural and epistemic knowledge of the disciplines (Kind & Osborne, 2017; Struyf et al., 2019), the absence of deep domain-specific knowledge is potentially frustrating for the learners and could result in the generation of low-quality solutions and quarrelsome debates that are not evidence-based.

Domain-specific knowledge required to solve a problem could be introduced to students before they start working on the problem. The different domain knowledge could be presented either through lectures, contextualised hands-on experiments, readings, or group discussions. The intention is for learners to be aware that the concepts are present and become familiar with them. For instance, if the students are tasked with building a portable water filtration system to purify water in a camp, the domain-specific knowledge could include knowledge of particle sizes and their filtration abilities (science), the concept of turbidity (science), factors affecting flowrate (science), how to measure and record flowrate (mathematics), and how differences in the diameter of the input and output could affect the pressure of water flow (engineering).

The domain-specific knowledge for meaningful engagement in the problem-solving process can be taught by different expert teachers using domain-specific strategies. For example, students can be tasked with designing experiments and investigating the relationship between particle size and flowrate in science, while the mathematics teacher could show students several worked examples of flowrate calculations to familiarise them with the variables involved in flowrate calculations. The engineering or design teacher can devise a demonstration using pipes to illustrate the resultant pressure differences from outlets with different diameters. Once students are aware and are familiar with the domain-specific knowledge associated with the problem, they can then proceed to decide which knowledge is required and apply it to design and production of a water purification system.

The development of domain-specific knowledge is a precursor for students to engage in constructing claims based on their understanding of the problem or the merits of their solutions. Correspondingly, domain-specific knowledge is also required for students to understand the claims made by others and raise relevant questions to critique them. In integrated STEM problem solving and inquiry, critique and claims related to success criteria are as important as debates based on domain-specific knowledge. For students who are novices in problem solving and STEM inquiry, teachers can present them with sentence starters or sentence stems as guides to scaffold students' thought processes as they formulate claims and statements to critique ideas. Sentence starters or stems have been used in a wide range of disciplinary contexts such as mathematics (Buffington et al., 2017), online discussions (Avci, 2020), and even in academic writing (Deveci, 2019). Examples of sentence starters are: 'The first factor we consider is ....', 'The variables that were changed were ....', 'I noticed that ....', 'I agree with your solution because ....', 'Could you please explain the part on ....', 'I need a little more information on ....', and 'What were the evidence you used for ...?'. Some sentence starters are domain-specific (asking for evidence), while others are generic and can be used for the process of problem solving.

When planning and enacting integrated STEM lessons, some questions you can ask yourself include: (1) What domain knowledge do students need to meaningfully engage with the STEM inquiry process? (2) How can opportunities be created for students to present their claims and critique the claims of others? (3) What domain knowledge will students learn by working through the STEM inquiry activity? and (4) What strategies can be used to enable students to clarify and justify their success criteria?

3.1.3 Condition 3: Space for design, making, and investigation (material aspect)

A peculiar affordance of integrated STEM is its relationship with the material world. Science and engineering are two disciplines that rely heavily on concrete materials for investigation and modelling. Creating time and space for students to work with three-dimensional designs, build physical

Disciplinarity in integrated STEM inquiry 51

models or prototypes, and carry out experiments is an important part of successful STEM learning. Feedback is needed during the making or investigative process to enable students to refine and improve their solutions or prototypes since refinement, improvement, and testing are important STEM inquiry processes. During STEM inquiry, data (in the form of observations or empirical evidence) serves as powerful feedback to students as to whether their idea or prototype is working. The data also help direct students' attention to areas that require refinement. This form of feedback can be embedded in the activity; it comes from students' direct interaction with the resource materials and is fast, accurate, and often unambiguous.

While students work with materials or apparatus that may not be familiar, feedback or advice from knowledgeable others allows them to be responsive to changes that are needed. For instance, science teachers can give students feedback in the following areas: (1) initiating an investigation, (2) conducting an investigation, or (3) presenting an investigation (So et al., 2018). For mathematics, feedback can be given during (1) data collection, (2) data processing, and (3) data representation. Peer feedback based on these well-established criteria is also a powerful and needful strategy to engage students in giving feedback and develop their 21st-century competencies in integrated STEM learning. Teachers can help students learn to take control of their own learning by defining learning goals and monitoring their progress in achieving them (Struyf et al., 2019). For example, in an enacted lesson where students worked in different groups to develop a water filtration system, questions such as 'What is the ratio of fine sand to gravel for optimal filtration ability?' were raised. Guided by the inquiry questions, students set up different investigations with 1:1 ratio of sand:gravel, 2:1 ratio of sand:gravel, and 1:2 ratio of sand:gravel. Using a measuring cylinder, they poured 50 mL of water through each set-up and recorded the time it took for the filtrate to emerge. They repeated the experiment at least three times, each time ensuring that the time recorded remained consistent. They then examined the times recorded in the form of a table and plotted a graph. They noticed that the higher the ratio of fine sand used, the longer it took for the filtrate to emerge. This was followed by a search for a possible explanation related to particle sizes and the amount of air spaces between them, and the creation of a model to illustrate their explanation. Their model resembled that of a pin-ball machine whereby the closer the particles, the more obstacles the ball had to travel through compared to particles that were further apart. The group debated if the air spaces were accurately represented by the spaces between the obstacles. While some felt that the spaces in between the obstacles were too large to represent air spaces, others were persuaded that the pin-ball model was an accurate representation. The students also discussed the limitations of their model since the large spaces within the obstacles suggested that particles with a wide range of sizes could pass through. This example highlighted the important role of working with concrete materials to enable students to build representative models of specific phenomena.

Questions to consider when designing a space for students to design, make, and investigate include: (1) What are the resources/materials/apparatus needed for a student to engage with making a prototype/model or conducting an investigation? (2) How do I evaluate the success of students' prototypes or models? (3) Are there embedded evaluations that will allow students to know which stage(s) is/are not working? (4) Will the materials that are provided lead students to common solutions?

3.1.4 Condition 4: Modelling communities of STEM problem solvers (social aspect)

Problem solving and STEM inquiry require students to communicate their intentions, ideas, assumptions, and biases with peers and authoritative others, such as their teachers. The application of various domain-specific knowledge in a systematic, logical, and coherent manner will not be successful unless students learn to collaborate with one another to (1) raise questions, (2) use tools to investigate, (3) interpret data, (4) reason and craft explanations, (5) create explanatory models to represent how observed phenomena work, (6) solve problems, (7) craft argumentations, (8) search and use structures, (9) attend to precision, and (10) think computationally. Further, to enable learners to develop the competencies necessary for productive engagement and empowerment in the 21st-century society, learning experiences need to offer opportunities for students to practice the skills of (1) adaptability, (2) persuasive communication, (3) non-routine problem solving, (4) self-management, and (5) systems thinking (Bybee, 2009). The ability to work with others and understand the social norms involved in STEM problem-solving teams is not trivial, and efforts must be put in to intentionally develop these competencies.

Opportunities can be created for students to experience interactions with others during problem solving and inquiry process. These can be done either through explicit teaching of these skills or through modelling these skills in the specific context of STEM inquiry. Through group engagement in STEM inquiry processes, students can be positioned to learn and practice 21st-century competencies, which include critical thinking, communication, collaboration, creativity, and innovation (Singapore Ministry of Education, 2016; Ontario Ministry of Education, 2016). STEM inquiry processes afford many opportunities for students to think critically, communicate, collaborate, and innovate as they apply domain-specific knowledge to novel situations. For instance, through explaining the reasoning behind their ideas, students develop their

Table 3.2 Mapping conditions of successful STEM inquiry to centricities

Conditions	Problem-centric STEM inquiry	Solution-centric STEM inquiry	User-centric STEM inquiry
Connections to real-world problems	• Make connections to contemporary issues within specific local contexts. • Problems are presented as complex, persistent, and extended.	• Consider interactions between students and the objects around them. • Present existing artefacts, technologies, or solutions to problems that are currently used in society.	• Interaction with members of the community to understand their experiences and diagnose the problems they face.
Opportunities for presentation and critique of ideas and claims	• Apply domain knowledge related to science and mathematics to state and critique claims. • Critique also involves understanding the context in which the problem is located.	• Apply domain knowledge related to engineering and technology to critique design and workability. • Knowledge is also used to understand affordances, strengths, and limitations of current designs/solutions/ways of doing things.	• Apply knowledge of user needs and current social context(s) to make claims and critique a product. • Compare existing policies and rules to the desired success criteria and user expectations.
Space for design, making, and investigation	• Students choose the materials they find most appropriate to design solutions. • Students test solutions by carrying out experiments and use the data to re-examine the problem and the solutions.	• Students work with existing prototypes/models to understand how they work and where changes/adaptations can be made. • Students test out their improved design against success criteria.	• Students consider ideas such as implementation details, accessibility, evaluation, and feedback mechanisms when users use the product/solution.
Modelling communities of STEM problem solvers	• Students work with experts (such as scientists or mathematicians) to learn techniques required to create solutions to the problems.	• Students work with industrial partners or suppliers to understand existing product features, intentions, and limitations.	• Students work with users in the community to understand different needs to make adaptations on the ground. • Emphasis is on impact authenticity.

Disciplinarity in integrated STEM inquiry 53

communication and critical thinking skills as they try to persuade their peers of their ideas and evaluate those of others. Through problem solving and computational thinking, students also practice and sharpen their competencies in creativity and innovation. Research in 21st-century competencies (Larson & Miller, 2012) underscores the value of meaningful, authentic, and contextually relevant experiences for students to formulate new experiences to sharpen their 21st-century competencies. Some questions to consider when deciding how to foster connections between domain-specific competencies and practices and 21st-century competencies are: (1) Are the 21st-century competencies intentionally mapped to the domain-specific learning outcomes? and (2) How can domain-specific knowledge and STEM inquiry practices connect?

3.2 Variations of integrated STEM inquiry

Variations of integrated STEM inquiry between different centricities may be considered for each condition as shown in Table 3.2. Teachers can consider the various adaptations for each condition and implement integrated STEM inquiry more flexibly, taking the conditions of actual classrooms and student profiles into consideration. These variations can also be interpreted as learners' expected behaviours and used as the basis for deciding which variant of the STEM Quartet framework should be adopted.

In Chapters 6 to 10, we present examples of how each condition is interpreted and enacted for different variants of the STEM Quartet instructional framework.

Questions for Reflection

1 How realistic are the conditions of successful integrated STEM inquiry for your professional practice? Which condition would be most challenging to fulfil?
2 What is the nature of the problems (referent-centred, problem-centred, or wicked-centred) you present to your students?

References

Avci, U. (2020). Examining the role of sentence openers, role assignment scaffolds and self-determination in collaborative knowledge building. *Educational Technology Research and Development, 68*, 109–135.
Bereiter, C. (1992). Referent-centred and problem-centred knowledge: Elements of an educational epistemology. *Interchange, 23*(4), 337–361.

Buchanan, R. (1992). Wicked problems in design thinking. *Design Issues, 8*(2), 5–21.
Buffington, P., Knight, T., & Tierney-Fife, P. (2017). Supporting mathematics discourse with sentence starters & sentence frames. *Interactive Technologies in STEM teaching and learning.*
Bybee, R. W. (2009). *The BSCS 5E instructional model and 21st century skills.* BSCS.
Deveci, İ. (2019). Reflections with regard to entrepreneurial project (E-STEM) process on the life skills of prospective science teachers: A qualitative study. *Journal of Individual Differences in Education, 1*(1), 14–29.
Donovan, M. S., Bransford, J. D., & Pellegrino, J. W. (1999). *How people learn: Bridging research and practice.* National Academy Press.
Driver, R., Squires, A., Duck, P., & Wood-Robinson, V. (1994). *Making sense of secondary sciences: Research into children's ideas.* Routledge.
Ford, M. J., & Forman, E. A. (2006). Redefining disciplinary learning in classroom contexts. *Review of Research in Education, 30*, 1–32.
Isik-Ercan, Z. (2020). 'You have 25 kids playing around!': Learning to implement inquiry-based science learning in an urban second-grade classroom. *International Journal of Science Education, 42*(3), 329–349. https://doi.org/10.1080/0950 0693.2019.1710874
Jonassen, D. H. (1997). Instructional design models for well-structured and ill-structured problem-solving learning outcomes. *Educational Technology, Research and Development, 45*(1), 65–94.
Kind, P. E. R., & Osborne, J. (2017). Styles of scientific reasoning: A cultural rationale for science education?. *Science education, 101*(1), 8–31.
Krajcik, J. (2015). Three-dimensional instruction. *The Science Teacher, 82*(8), 50.
Larson, L. C., & Miller, T. N. (2012). 21st century skills: Prepare students for the future. *Kappa Delta Pi Record, 47*, 121–123. https://doi.org/10.1080/00228958.2 011.10516475
Levy, F., & Murnane, R. (2004). *The new division of labor: How computers are creating the next job market.* Princeton University Press.
Ministry of Education, Singapore. (2016). *21st century competencies.* Retrieved from https://www.moe.gov.sg/education/education-system/21st-century-competencies
Ministry of Ontario Education. (2016). *Phase 1- Towards defining 21st century competencies for Ontario.* Retrieved from https://www.kslaring.no/pluginfile. php/57624/mod_page/content/1/21stCentury%20Competencies.pdf
National Research Council [NRC]. (2000). *Inquiry and the national science education standards.* National Academy Press.
Ong, Y. S., Koh, J., Tan, A. L., & Ng, Y. S. (2023). Developing an integrated STEM classroom observation protocol using the productive disciplinary engagement framework. *Research in Science Education*, 1–18. https://doi.org/10.1007/ s11165-023-10110-z
Priemer, B., Eilerts, K., Filler, A., Pinkwart, N., Rosken-Winter, B., Tiemann, R., & Belzen, A.U. Z. (2020). *A framework to foster problem-solving in STEM and computing education.* Research in Science & Technological Education, 38(1), 105–130, https://doi.org/10.1080/02635143.2019.1600490
Rittel, H. W. J., & Webber, M. M. (1973). Dilemmas in a general theory of planning. *Policy Sciences, 4*(2), 155–169.

Roach, K., Tilley, E., & Mitchell, J. (2018). How authentic does authentic learning have to be? *Higher Education Pedagogies*, *3*(1), 495–509. https://doi.org/10.1080/23752696.2018.1462099

Sias, C. M., Nadelson, L. S., Juth, S. M., & Seifert, A. L. (2017). The best laid plans: Educational innovation in elementary teacher generated integrated STEM lesson plans. *The Journal of Educational Research*, *110*(3), 227–238. https://doi.org/10.1080/00220671.2016.1253539

Sinatra, G. M., & Pintrich, P. R. (2002). The role of intentions in conceptual change learning. In G. M. Sinatra & P. R. Pintrich (Eds.), *Intentional conceptual change* (pp. 1–17). Routledge.

Struyf, A., De Loof, H., Boeve-de Pauw, J., & Van Petegem, P. (2019). Students' engagement in different STEM learning environments: integrated STEM education as promising practice? *International Journal of Science Education*, *41*(10), 1387–1407.

United Nations (2015). About the sustainable development goals. Retrieved from https://www.un.org/sustainabledevelopment/sustainable-development-goals/

Wade, A. A., Grant, A., Karasaki, S., Smoak, R., Cwiertny, D., Wilcox, C., Yung, L., Sleeper, K., & Anandhi, A. (2020). Developing leaders to tackle wicked problems at the nexus of food energy, and water systems. *Elementa Science of the Anthropocene*, *8*(1), 1–13, https://doi.org/10.1525/elementa.407

4 Assessing integrated STEM inquiry

Assessment is an integral part of teaching and learning. In integrated STEM inquiry, we also have to consider how students' learning can be assessed. Assessment in integrated STEM inquiry is unlike traditional forms of assessment, where students' attainment of conceptual understanding is measured through robust paper-and-pencil tests or where students' ability to carry out experiments is determined by performance tasks. In integrated STEM learning, assessment is multi-faceted and multi-level, taking the form of a suite of different assessment methods that are dependent on the nature of the activities that students engage with in the different phases of inquiry. In this chapter, we discuss the various assessment models for integrated STEM inquiry with suggestions for how each can be implemented.

4.1 An assessment model for integrated STEM inquiry

What do you think is the most important aspect of assessment for STEM learning? Given that STEM learning is oriented to students' experiences of real-world problem solving by integrating various disciplinary understandings, skills, and competencies, STEM assessment also needs to reflect this fundamental philosophy. However, Gao et al. (2020) pointed out that the assessments of existing STEM programmes differ from the philosophy of STEM in two ways.

First, it was reported that most existing STEM programmes tend to assess students' monodisciplinary understanding or skills despite intentions to foster interdisciplinary practices (Gao et al., 2020). The assessment of monodisciplinary knowledge, skills, or affective aspects was most common, whereas the assessment of interdisciplinary or transdisciplinary knowledge and practices was much less. This tendency of assessment in STEM programmes results in inconsistencies in the objectives, instruction, and assessment in the STEM programme.

Second, most STEM programmes focus on assessing the final product(s) of their STEM inquiry (Gao et al., 2020). This is problematic because assessing only the final product(s) does not consider the iterative process of modelling, designing, getting feedback, and refining the product, which is an important practice in STEM (Namdar & Shen, 2015). The process of working with materials and peers in the creation of a solution is a necessary condition for successful STEM inquiry. The iterative process of modelling, designing, and refining allows students to improve their skills in creativity, critical thinking, and collaborative communication. A final product-focused assessment can prevent students from developing these skills.

Assessments that truly reflect the philosophy and intent of integrated STEM learning should focus on a specific activity's context and the processes in which transdisciplinary knowledge and skills are naturally integrated during STEM practices, rather than assessing only the final product. For example, we can assess student's competency of collaborative problem solving in the context of STEM practices in their group activity (Herro et al., 2017). Assessment for integrated STEM learning is intentional and specific and takes into account the relationships among STEM disciplines within the curriculum (Bryan et al., 2015). Given this, we suggest two key ideas for integrated STEM assessment in this chapter: (1) an integrated competence-based assessment and (2) a suite of different assessment methods depending on the STEM lesson phase.

4.1.1 An integrated competence-based assessment

Well-designed integrated STEM inquiry should provide opportunities for students to connect and integrate STEM disciplinary concepts and construct various inferences and models to create solutions by combining their everyday experiences, knowledge, and high-level thinking skills, such as problem-solving skills or 21st-century competencies. Aligned with the intention of integrated STEM inquiry, an integrated STEM assessment needs to assess the competencies students require in the following integrated three areas.

Ideally, this model suggests that student competencies ought to be assessed in an integrated form within a specific STEM activity, rather than simply stacking domain-specific competencies. Once the transdisciplinary connections are clarified in STEM curricula and instruction, it is recommended that these interdisciplinary connections are captured and assessed in students' actual STEM learning processes (Gao et al., 2020). The practical ways of integrating several competencies in the three proposed areas and deciding on the categories of assessment will be discussed in detail in Section 4.2.

4.1.2 An integrated suite of assessment methods

Developing well-crafted assessment methods and tools for classroom use can provide both teachers and students with high-quality information about students' learning and the effects of instruction (Fang & Hsu, 2019). The assessment criteria for the activities in an integrated STEM inquiry task can first be determined before we move forward with the development of specific assessment methods and tools.

When developing assessment methods, it is recommended that assessment tools and rubrics be crafted by linking them to the specific phases and levels of a STEM lesson. For instance, teachers can plan how to assess student learning at various levels, such as individual, group, or class levels depending on the objectives of the lesson phase. Since students can engage with STEM activities either individually or collaboratively during STEM inquiry (Donmez, 2020), STEM assessment needs to be designed with a consideration of the specific context of the lessons. For instance, it is helpful to establish rubrics or protocols to assess students' competencies because rubrics can provide a clear description of the criteria teachers will use to assess student learning (Ong et al., 2023). We will focus more on assessment methods specified for integrated STEM inquiry in Section 4.3.

In integrated STEM inquiry, assessment can take the form of an integrated suite of different assessment categories and methods depending on (1) the lesson intention(s) and (2) the phases of integrated STEM inquiry in which students engage. Based on the two ideas of 'an integrated competence-based assessment' and 'an integrated suite of assessment methods', the following sections will describe two steps of STEM inquiry assessment with practical examples. In the first step, *what* can be assessed in integrated STEM learning will be discussed. In this step, we discuss how decision on the different assessment categories of integrated competencies can be made. In the second step, we describe, in terms of the assessment method, *how* various forms of assessment tools or rubrics specific to an activity can be developed.

4.2 Assessing *what*: Deciding on the categories of an assessment

Multiple components have been assessed in STEM programmes (Gao et al., 2020) as STEM education pursues broad goals and values (Teo et al., 2021). Fang and Hsu (2019) suggested different categories of assessment addressing multi-faceted learning achievements in STEM, such as disciplinary or integrated knowledge, competencies, literacy, and attitudes. Table 4.1 presents three categories and their subcomponents that can be assessed in STEM learning.

60 *Assessing integrated STEM inquiry*

Table 4.1 Three categories of assessment in STEM learning

Authors (Year)	Categories of assessment in STEM learning		
	Conceptual knowledge	*Competencies/practices/ skills*	*Attitude*
Ryu et al. (2019)		Creative thinking, critical thinking, collaborative problem solving, and communication	
Fang & Hsu (2019)	Knowledge Integration	Competencies (e.g., problem-solving abilities, computational thinking skills) and literacy (e.g., technology and engineering literacy)	Attitude (e.g., career interests)
Gao et al., (2020)	Understanding	Skills	Affective aspect(s)
Cheng & Yeh (2022)	Knowledge	Practices and skills	Attitude

Rather than an assessment of knowledge itself (see Figure 4.1), the assessment of integrated STEM inquiry needs to align with the philosophy of integrated STEM learning and focus on competency-based assessment. Table 4.1 illustrates the diverse sub-components of competency that have been discussed in detail and variety so far. One can assess, within each category of competency, a student's creativity, critical thinking, computational thinking, problem-solving abilities, communication skills,

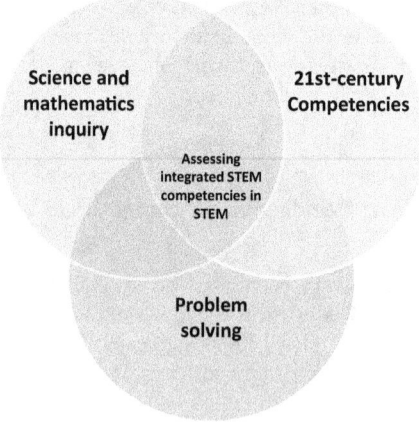

Figure 4.1 A competence-based assessment integrating three key areas.

collaborative skills, and literacies during their STEM practices. These competencies can be assessed in an integrative manner within a specific activity as discussed in Figure 4.1.

What does integrated forms of competencies mean? Here, the integrated forms of competencies refer to the contextualised competencies that are required to solve a specific STEM problem. Table 4.2 shows examples of integrated forms of competencies contextualised for different STEM problems.

Each problem can have a different variation of STEM competencies depending on the related disciplinary knowledge, skills, structuredness, and complexity (Jonassen, 2000). Depending on the STEM problem, different competencies are required of students in different activities. For example, in Table 4.2, the building waterpipes project requires student competencies in inquiring about the relationships between the geological characteristics of the mountain areas and the consequent water pressure

Table 4.2 Examples of integrated forms of competencies within the context of STEM problems

STEM problem	→	Interpreting competencies in an integrated form within a STEM problem	→	Extracting key competencies for assessment
Designing water purification devices	→	- Inquiring about the chemical characteristics of water quality. - Designing an engineering device for water purification.	→	- Ask science-oriented questions. - Analyse data to show trends. - Assess product creativity through novel ways of using materials to build the devices.
Building waterpipes in the mountains	→	- Inquiring about the relationships between the geological features of the mountain area and the consequent water pressure. - Creating technology-based representations of mountains and water pipes based on graph theory.	→	- Analyse data to describe relationships. - Use cognitive creativity to create a representation of ideas. - Make inference(s) based on observations and data.

and constructing technology-based models. On the other hand, designing water purification devices for the school canteen project requires other competencies including inquiring about the chemical characteristics of water quality and designing and creating an engineering product. As such, the student competencies to be assessed can vary depending on the nature of the STEM problem.

After student competencies have been interpreted in an integrative manner within the context of a STEM problem, teachers can extract key competencies fit for assessing each activity effectively. Table 4.3 shows how to design and use a rubric that helps teachers determine assessment categories focused on competence-based assessment in integrated STEM inquiry. As shown in this rubric, practical instruments can be used to determine the categories for STEM assessment.

The competencies presented in Table 4.3 are general competencies utilised in integrated STEM inquiry. However, these competencies may vary depending on the context of the STEM activity. In STEM activity design and assessment, the competencies extracted should be tuned to the STEM activity designed.

In this way, integrated competency-based assessment focuses on the competencies that incorporate the interdisciplinary knowledge and skills

Table 4.3 Using a rubric to determine assessment categories in integrated STEM inquiry

Categories for competence-based assessment in integrated STEM inquiry						
Asking questions	◻	Scientifically oriented questions	◻	Procedural questions	◻	Practical questions
Data analysis	◻	To identify cause and effect	■	To describe relationships	◻	To show trends
Creativity	■	Cognitive creativity	◻	Process creativity	◻	Product creativity
Critical thinking	◻	Rationale-based inference	■	Inferences based on sequence of events	◻	System-based inference
Collaborative problem solving	◻	Interest sharing	■	Role sharing	◻	Collaborative Monitoring
Communication	◻	Description of ideas and observations	■	Expressions of opinion	◻	Productive criticism

required in STEM activities. Competency-based assessment aligns with the goal of integrated STEM learning as it moves away from existing methods of assessing monodisciplinary competencies as separate components, promoting interdisciplinary integration.

4.3 Assessing *how*: Developing a suite of different assessment methods

Determining the assessment method used entails considering various factors, such as assessment context, instruments used, and types of feedback (Chappuis et al., 2012). Careful consideration is needed in terms of determining 'in which phase', by 'who[m]', and at 'which level' should the assessment be carried out.

4.3.1 In which phase? (phases of STEM inquiry in STEM assessment)

Integrated STEM inquiry processes entail investigating related STEM disciplinary information/data/knowledge to solve an authentic problem and designing solutions embodied in the forms of a sketch/drawing, scaled/working model, or actual object. These processes can be categorised into three phases of STEM: (1) *problem definition*, (2) *research*, and (3) *development* (Ong et al., 2023).

In the problem definition phase, the context of the task is introduced, and the problems related to the context are clearly defined. In the research phase, students are required to search for relevant existing knowledge/information and carry out investigations to gather data. Lastly, in the development phase, students generate ideas/representations of their proposed solutions, test, and revise or improve their solutions/prototypes. It is important in this last phase to entail the iterative process of testing, reviewing, and revising their solutions/prototype. A suite of different assessment methods can be developed and organised based on the main focus of the three phases.

4.3.2 By whom? (agents of STEM inquiry assessment)

In STEM assessment, there can be various agents who can report on students' learning. The assessment methods can be designed and implemented differently depending on the assessor. Specifically, assessment of STEM learning can be categorised as *student self-reported* or *peer-reported*, *teacher-subjective*, or *objective* (Cheng & Yeh, 2022).

Student self-report or peer-report assessments involve having students assess their own works/activities or the works/activities of their peers, for instance, in peer-group to peer-group assessments, where one group provides critical feedback to another group and recommends ways that the other group could improve on their creations. Space is also created for the group being assessed to defend their solution or acknowledge the feedback with a view to tasking on board the feedback.

These methods can help students develop metacognitive skills and a better understanding of the criteria used in evaluation. However, they can also be influenced by a student's self-perception or personal feelings towards their peers. Teacher-subjective assessment can include methods such as observational assessment, where teachers note behaviours or skills as they observe students in the classroom or grade an essay based on a holistic impression of its quality. These assessments can be sensitive to aspects of performance that are difficult to quantify and can require a deep understanding of each learner.

Meanwhile, objective and standardised tests are designed to measure students' understanding of specific content and are easy to grade consistently. The results obtained can be compared across different students, classes, schools, and even states and countries, providing valuable data for evaluating educational policies and practices. However, these tests may not fully capture a student's abilities or potential, especially when it comes to skills like creativity, critical thinking, or teamwork, which are harder to measure objectively. Figure 4.2. shows an overview of the three assessment methods used to assess student learning domains in STEM education so far.

As reported in Figure 4.2, the evaluation method most commonly used in STEM evaluation is student self-reported information. Student self-reported assessment usually covers a whole suite of STEM assessment, such as knowledge, skills, and attitude. In comparison, teacher-subjective assessment and objective assessment are less commonly used. Teacher-subjective assessment tends to cover three assessment domains, whereas objective assessment primarily targets evaluation of students' knowledge and practices.

Depending on what is being assessed, multiple agents may be required for holistic competence-based assessment in integrated STEM learning. As shown in Figure 4.2, objective assessment was only used to assess

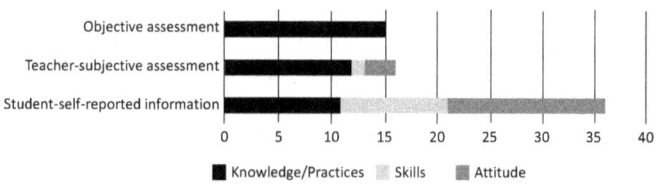

Figure 4.2 Assessment methods used in STEM education to evaluate student learning domains (Cheung & Yeh, 2022).

Assessing integrated STEM inquiry 65

students' knowledge and practice. However, standardised objective assessments are limited in assessing students' skills and competencies, constructs that are emphasised in integrated STEM inquiry. Thus, when evaluating students' skills and attitudes, teacher-subjective, and student self-reported assessments are more suitable methods of competence-based assessment in integrated STEM inquiry.

4.3.3 At what level? (levels of STEM assessment)

Carrying out a collaborative group project is a common way of engaging in a STEM learning activity (Herro & Quigley, 2016). From an assessment perspective, the learning context of a group project provides opportune moments to observe and capture how student competencies are actualised and developed. For example, group discussions in project-based activities can provide a chance to assess how students interact with their peers collaboratively and positively (Herro et al., 2017). When students have group presentations to share their project's progress, we can check how student groups connect and integrate STEM disciplinary knowledge into their problems and communicate their ideas effectively. In this way, different methods of assessment can be set up to identify and check student competencies during group projects.

Various assessment contexts can be implemented at the individual and group levels for group-based projects. Tables 4.4 and 4.5 present examples of rubrics for student self-reported and peer-reported assessments at the individual and group levels, respectively.

The rubric presented in Table 4.4 allows students to self-assess their contributions to their group's practices. This self-assessment of (individual contribution) in group projects can offer a chance for students to engage in personal reflection as they reflect on their participation, collaboration, and attitude. On the other hand, in Table 4.5, students were asked to assess their peers as a group and evaluate how each member

Table 4.4 An example of rubric for individual student self-reported assessment

I contributed to my team in the following ways:

Ideas	Organisation
▫ I contributed at least one idea. ▫ I provided feedback to the ideas presented by others in a respectful manner. ▫ I listened attentively to ideas presented by others without interrupting.	▫ I gave at least one feedback when we were setting the goals for the activity. ▫ I worked with others in the group to craft the timeline for the activity.

Table 4.5 An example of rubric for group peer-reported assessment

	Student A	Student B	Student C
Contribution to ideas	☺	☺☺	☺
Listening to others	☺		☺☺
Providing feedback		☺☺	

contributed to their group's practices. Depending on their personal preferences, competence and/or interests, each member would contribute to their group project differently. Thus, students can check the overall participation and contribution of their members at a group level.

Teachers also play an important role in evaluating student performance in group project-based STEM learning. Their observations, garnered from guiding project-based activities, can be an invaluable resource for the monitoring and assessment of students' practices. Table 4.6 demonstrates one method of establishing a concrete rubric for teachers to assess project-based STEM practices, providing score-based criteria for group evaluation.

The different levels of students' practices may be observed in the same phase; however, they need to be assessed differently. As shown in Table 4.6, in the problem-solving phase, students who generate creative and sustainable solutions can be evaluated differently from those who provide typical solutions.

In conclusion, to effectively construct and implement assessment methods in integrated STEM learning, we must carefully consider three aspects: the phase of assessment, the assessment agents, and the level of assessment. From these three perspectives of assessment, we can

Table 4.6 An example of rubric for teacher-subjective group assessment

Phase	Score			
	1	2	3	4
Problem-solving phase	Typical solution that resembles the current way of doing things	Thoughtful solution that is sustainable only for a short period but does not have growth potential	Thoughtful solution that is sustainable only for a short period but has growth potential and challenges existing ways of thinking	Thoughtful elegant solution that is sustainable and has growth potential and challenges existing ways of thinking

Assessing integrated STEM inquiry   67

Table 4.7 Categorising and developing a suite of different STEM assessment methods

In which phases?	By whom?	At which level?
■ Problem definition phase	□ Student self-reported assessment ■ Peer-reported assessment □ Teacher-subjective assessment □ Objective assessment	■ Individual-level assessment □ Group-level assessment
■ Research phase	□ Student self-assessment □ Peer-reported assessment ■ Teacher-subjective assessment □ Objective assessment	□ Individual-level assessment ■ Group-level assessment
□ Development phase	□ Student self-assessment □ Peer-reported assessment □ Teacher-subjective assessment □ Objective assessment	□ Individual-level assessment □ Group-level assessment

categorise and develop a suite of different STEM assessment methods as shown in Table 4.7. Table 4.7 shows a rubric for categorising and planning assessments—it does not contain the assessments themselves. Rather, this rubric offers a way to systematise and visualise how different assessment strategies align with the various phases of the integrated STEM learning process, by whom the assessment is carried out, and at which level should the assessment be conducted.

Let's take a brief look at each column of the rubric. The first column, 'In which phase?', pertains to the stages in an integrated STEM learning process during which the assessment methods are implemented. Next, 'By whom?' designates the persons or groups responsible for executing the assessment(s). Lastly, 'At which level?' corresponds to the level of the assessment being undertaken. Given its design, this rubric can afford teachers a systematic way to construct and outline a different suite of assessment strategies for different integrated STEM activities.

4.4 A formative and iterative process of assessing integrated STEM inquiry

To consolidate the key points from this chapter, we propose a formative and iterative framework of assessment as a way of promoting student STEM inquiry. The categories and methods/tools of assessment can be

68 *Assessing integrated STEM inquiry*

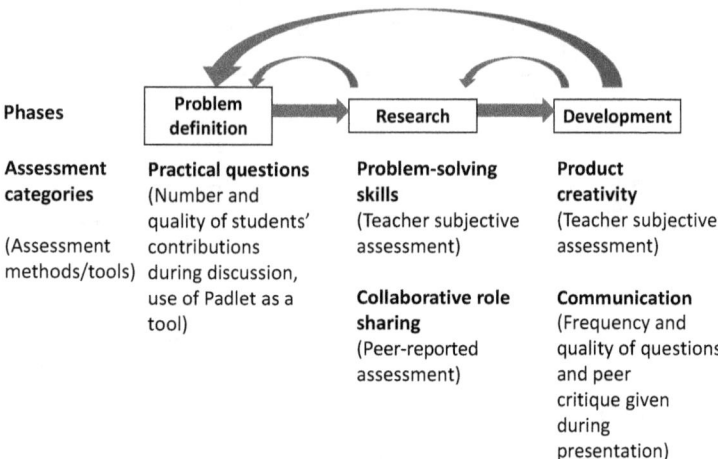

Figure 4.3 A formative and iterative process of STEM practice assessment.

designed and implemented differently depending on the phases of integrated STEM inquiry. Figure 4.3 illustrates how to create a suite of diverse assessment methods or tools tailored for each phase of STEM practice.

The key point of the above process is that the assessments in each phase are interrelated and iterative. This is because the practices of integrated STEM inquiry are iterative in nature. Just as students create their solutions through several iterations of defining and redefining the problem, researching, testing, and revising ideas, teachers also need to monitor, check, and assess students' practices in a similar iterative way.

Suppose a teacher realises that students in the research phase are unable to generate any ideas to problem solve, this could indicate that the problem definition in phase 1 was too broad or vague or that assistance is needed to connect related disciplinary knowledge/skills during their problem-solving process. If students' problems are too broad, the teacher can guide them to narrow down their problem by (re)checking and (re)assessing the quality of the problems defined. Should the students have difficulties in applying disciplinary knowledge/skills in their problem solving, the teacher can demonstrate specific methods that can be applied or examples of their application. The subsequent vignette illustrates how a teacher can facilitate students' problem-solving processes and assess their problem-solving skills through formative assessment (Figure 4.4).

Figure 4.4 An example of teacher-subjective assessment of problem-solving skills.

Phase	Score		
	1	2	3
Problem-solving skills	Failed to incorporate the turbidity sensor and the concept of flowrate of water in the system design	Partially applied incorporated the turbidity sensor and the concept of flowrate in the system design	Effectively incorporated the turbidity sensor and the concept of flowrate in the system design

---

Water treatment vignette: The research phase in a STEM lesson on designing a water treatment system.

The students constructed their codes (very cloudy, cloudy, and clear) and calibrated their turbidity sensors using water that was collected earlier by the teachers. Following the testing of the sensors, another computer teacher took over the class and introduced the students to the concept of flowrate, which can be used to evaluate students' design of their wastewater system

---

In the case above, the teacher was able to guide the students on how to apply disciplinary skills in their problem solving and assessed their problem-solving skills and practices at the same time.

Similarly, in the development phase, the iterative processes of researching, testing, and revising of ideas can be fostered and assessed through formative assessment. In the following vignette, students collaborated within and across groups to critically evaluate the feasibility of their prototypes as a pigeon repellent. The students assessed their peers' collaboration through a peer-reported assessment as shown in Figure 4.5.

Pigeon repellent vignette: The development phase in a STEM lesson on pigeon repellent. The feedback pointed out actual possible critical and creative issues. The comments regarding the ultrasonic waves included 'the equipment won't disturb humans', 'the equipment will be protected by rain', and 'where to install?'. Based on meaningful collaboration and communication, the students were able to research additional related information and revise the design of their model of pigeon repellent.

Figure 4.5 An example of peer-reported assessment of collaboration in a round-robin activity.

	Student A	Student B	Student C
Researching and suggesting ideas	☺	☺☺	☺
Listening to others	☺		☺☺
Providing critical feedback		☺☺	

From an assessment perspective, the above collaborative activity can provide a good opportunity to assess student collaboration in their integrated STEM practices. This is an example of focusing on the social aspect of STEM inquiry. The round-robin activity above allowed the students to investigate additional related issues, such as the impact(s) on humans, rainproofing abilities, and suitable installation location. Based on the additional research, the students constructed prototypes of their models refining their designs to further elaborate on and develop their solutions. As such, the activity featured in the vignette can create an environment for formative assessment by revealing students' contributions to the ideation process and their collaborative attitudes.

There is no set path to assessing STEM learning; there are many ways to assess students' STEM practices, depending on learner interest, level, group size, classroom environment, activity/topic, and more. That is why, rather than relying on generic assessment resources, it is important for teachers to possess the expertise to determine the best assessment categories and methods suitable for the contexts of their integrated STEM lessons.

> **Questions for Reflection**
>
> 1 How are assessment and evaluation in STEM inquiry different from other forms of assessment and evaluation?
> 2 Describe some strategies that can be used to evaluate 21st-century competencies.
> 3 Why is it important to align assessment modes/methods with the intended learning outcomes of STEM inquiry?

References

Bryan, L. A., Moore, T. J., Johnson, C. C., & Roehrig, G. H. (2015). Integrated STEM education. In C. C. Johnson, E. E. Peters-Burton, & T. J. Moore (Eds.),

STEM road map: A framework for integrated STEM education (pp. 23–38). Routledge.
Chappuis, J., Stiggins, R. J., Chappuis, S., & Arter, J. (2012). *Classroom assessment for student learning: Doing it right-using it well* (p. 432). Pearson.
Cheng, M. M. H., & Yeh, F. Y. (2022). Identifying effective STEM programmes and strategies in Asia. In M. M. H. Cheng, C. Buntting, & A. Jones. (Eds.), *Concepts and practices of STEM education in Asia* (pp. 19–41). Springer Nature Singapore.
Donmez, I. (2020). STEM education dimensions: from STEM literacy to STEM assessment. In S. Idin (Ed.), *Research highlights in education and science* (pp. 154–170). International Society for Research in Education and Science.
Fang, S. C., & Hsu, Y. S. (2019). Assessment challenges in STEM reforms and innovations. In Y. S. Hsu & Y. F. Yeh (Eds.), *Asia-Pacific STEM teaching practices* (pp. 191–203). Springer.
Gao, X., Li, P., Shen, J, & Sun, H. (2020). Reviewing assessment of student learning in interdisciplinary STEM education. *International Journal of STEM Education*, 7, 24. https://doi.org/10.1186/s40594-020-00225-4
Herro, D., & Quigley, C. (2016). STEAM enacted: A case study of a middle school teacher implementing STEAM instructional practices. *Journal of Computers in Mathematics and Science Teaching*, 35(4), 319–342.
Herro, D., Quigley, C., Andrews, J., & Delacruz, G. (2017). Co-measure: Developing an assessment for student collaboration in STEAM activities. *International Journal of STEM Education*, 4, 1–12.
Jonassen, D. H. (2000). Toward a design theory of problem solving. *Educational Technology Research and Development*, 48(4), 63–85.
Namdar, B., & Shen, J. (2015). Modeling-oriented assessment in K-12 science education: A synthesis of research from 1980 to 2013 and new directions. *International Journal of Science Education*, 37(7), 993–1023.
Ong, Y. S., Koh, J., Tan, A. L., & Ng, Y. S. (2023). Developing an integrated STEM classroom observation protocol using the productive disciplinary engagement framework. *Research in Science Education*. https://doi.org/10.1007/s11165-023-10110-z
Ryu, M., Mentzer, N., & Knobloch, N. (2019). Preservice teachers' experiences of STEM integration: Challenges and implications for integrated STEM teacher preparation. *International Journal of Technology and Design Education*, 29, 493–512.
Teo, T. W., Tan, A. L., Ong, Y. S., & Choy, B. H. (2021). Centricities of STEM curriculum frameworks: Variations of the STEM Quartet. *STEM Education*, 1(3), 141–156.

5 Enactment of integrated STEM inquiry in classrooms

The first three chapters form the theoretical foundations for the interpretation of integrated STEM inquiry in the classroom, while Chapter 4 discussed issues related to assessment of students' learning. In Chapter 5, we revisit the typical characteristics of the three STEM Quartet variants, which will serve as a frame for interpreting classroom practices from Chapters 6 to 10.

5.1 Flexibility in integrated STEM inquiry enactment

As described in Chapter 3, the four essential conditions for successful integrated STEM inquiry that we need to bear in mind are:

1. Connection to real-world problems;
2. Opportunities for presentation and critique of ideas and claims (inter-play of role as Constructor and Critiquer of claims);
3. Space for design, making, and investigation (material aspect); and
4. Modelling communities of STEM problem solvers (social aspect).

These four conditions serve as planning frames to scaffold teachers' pedagogical decision making as they design and implement each of the three variants of integrated STEM inquiry.

Five vignettes of actual integrated STEM inquiry lessons are used to illustrate how the four conditions are infused for each variant and the pedagogical practices that are used to fulfil the planned learning intentions. The aim of the STEM lesson vignettes is to help readers visualise how each variant of the STEM Quartet framework can be actualised in STEM learning. Vignettes of enacted integrated STEM inquiry in the classroom provide insights into how teachers translate their ideas and plans into actual practices. We have intentionally chosen to use vignettes as they can capture the essence of a classroom situation and present what

DOI: 10.4324/9781003422501-5

is real, hence creating opportunities for discussion of the issues pertaining to teaching and learning of integrated STEM (Veal, 2002).

The five vignettes were created from classroom observation field notes recorded by researchers working with Thai teachers in their classrooms. These teachers were part of a collaborative project aimed at empowering teachers with practice knowledge related to integrated STEM. The teachers attended workshops on how to plan STEM learning experiences and worked with the researchers in a professional learning team to plan their lessons.

5.2 Conditions for integrated STEM inquiry enactment

5.2.1 Problem-centric STEM inquiry

When teachers begin their lessons by presenting a complex, persistent, and extended problem for their students to work on, they are adopting the problem-centric STEM inquiry variant. To engage students' interests and experiences in problem-centric STEM inquiry, students are guided to make connections to contemporary issues within specific local contexts. Intentionally connecting students' familiar experiences to a problem is important as issues, objects, or phenomena can become 'invisible' to the students due to familiarity with them (refer to Table 3.1, Social aspect of introducing the context in which the problem is situated).

Making observations and gathering data enable students to develop greater clarity of the extent and scope of the problem encountered (refer to Table 3.1, Clarifying the solution requirements and requesting for justification(s)). Problematising familiar experiences requires deliberate efforts to 'see' what has become 'invisible' in the daily experiences of students. Creating opportunities for students to connect their knowledge and skills to the practices of science and mathematics allows for evidence to be collected during an integrated STEM inquiry process. Further, to establish meaningful real-world problems, students should interact with complex, persistent, and extended problems that communities and the society at large have grappled with through time. Next, students require a deep understanding of domain-specific knowledge related to the problem in order to generate solutions and engage in meaningful discussions with their peers. That is, learners need to be guided to identify the relevant domain knowledge from science and mathematics that are useful to specific problems and present these ideas in a public space for critique. They should also be guided to apply the domain knowledge to generate plausible solutions. Sound domain knowledge and skills play key roles in enabling students' development of common conceptual understandings and language skills to engage in discussion and craft explanations as they problem solve. In the absence of deep disciplinary knowledge, no connections between disciplines or the problem can be made. Hence, the role of deep disciplinary knowledge cannot be undermined or overshadowed by a sole focus to develop problem-solving skills. In problem-centric STEM inquiry, the disciplinary knowledge of science and mathematics must be coupled with knowledge of problem solving and creating representations of solutions. The last condition for meaningful integrated STEM inquiry is engagement with other problem solvers in the community. During problem solving and STEM inquiry, feedback can take various forms such as oral feedback, using data collected to check for efficacies of systems developed, sequential flow systems to ensure that each step is accurately carried out, peer critique, etc. Multiple sources of feedback provided throughout the entire process of problem-solving offer students numerous opportunities to evaluate their ideas (refer to Table 3.1, Reflecting on the STEM learning experience).

While engaging in the STEM inquiry process, learners can engage in collaborative discussions and group critique and, in the process, develop their abilities to defend their ideas during the problem-solving process. The guiding questions provided by teachers allow students to reflect on their solutions in a collaborative manner so that multiple perspectives can be considered. Another important means of getting feedback is the use of data. Students investigate the effectiveness of their solutions by designing experiments and collecting data. The data gathered serves as objective feedback for each solution. In this way, learners consider the appropriateness of both the problem and corresponding solutions. Ultimately, activities carried out to solve problems in lessons planned with the problem-centric framework aim to focus students' attention on how domain knowledge and skills can be applied to making sense of problems they encounter in their daily lives. Students also engage in generating and testing possible solutions to the specific problems they encounter using thinking processes required for problem solving.

5.2.2 Solution-centric STEM inquiry

In solution-centric STEM inquiry, students are first presented with one or more existing solutions that are currently available for evaluation. Students are provided with opportunities to be more productive and broadly engage in the 'looking back' phase of problem solving, which is easily ignored in the problem-solving process (Teo et al., 2021). Learners can engage in an iterative process, starting with an existing solution, evaluating the solution, looking back on the gaps in the evaluated solution, and creating and testing improved solutions (refer to Table 3.1, Epistemic aspect of clarifying the solution requirements and requesting for justifications).

For the first condition in the solution-centric framework, learners are guided to consider interactions between themselves and objects/phenomena/ways of doing things around them (refer to Table 3.1, Conceptual/Procedural aspect of pacing and maintaining momentum of an iterative problem-solving cycle). This consideration is usually anchored on specific purposes or intentions. Similar to problem-centric inquiry, the context presented to students in solution-centric inquiry can be one that students are familiar with. However, unlike problem-centric inquiry, students are guided to develop dominant domain knowledge related to engineering and technology in solution-centric inquiry. To connect knowledge to practice, students are required to design technical products and develop creative experimental skills. Finally, the system for testing, feedback, and evaluation in solution-centric inquiry involves students testing, obtaining data, and refining solutions/artefacts for improvement; This iterative process is one of the key steps in solution-centric inquiry. Through the making, testing, and improvement of their prototypes, students need to be inventive and think critically when facing failures and making modifications. Furthermore, the iterative process in group work affords students opportunities to practice their skills of communication and collaboration with their teammates.

5.2.3 User-centric STEM inquiry

In user-centric STEM inquiry, students begin by seeking to understand user needs, preferences, and challenges. Students work on understanding the users and their contexts, envisioning and creating a solution for the users, and evaluating the solution from the users' perspective (refer to Table 3.1, Social aspect of introducing the context and problem). For the first condition in user-centric STEM inquiry, students are guided to create opportunities to listen, understand, and distil problems faced by users of specific technologies, artefacts, or solutions. Second, the focus in this form of inquiry is on understanding the users' problems in the local context where the problem is situated. It is worth noting that students may not be accustomed to considering the user's point of view, therefore providing feedback in various ways is recommended. For the third condition in the user-centric framework, students are guided to focus on how domain-specific knowledge in S, T, E, and M interacts with the users' problems.

5.3 Hybrid applications of implementing integrated STEM inquiry

The three variants of the STEM Quartet instructional framework can be implemented in a hybrid way in the classroom. Earlier, we discussed and highlighted the typical characteristics and nuances of each of the three variants (i.e., problem-centric, solution-centric, and user-centric inquiry) separately. However, within the context of an actual dynamic classroom, a STEM lesson can be planned with two or more variants flexibly applied, depending on the intention of each activity in the STEM lesson. This is because STEM lesson objectives, problems, and classroom situations vary widely but are deeply interconnected with each other.

Table 5.1 shows a concrete example of a hybrid application of integrated STEM inquiry variants. The example shows how the conditions of different variants can be adapted in one STEM lesson. As shown in Table 5.1, the elements of the problem-centric variant were selected for the first and second conditions, whereas the elements of the user-centric variant were selected for the third and fourth conditions.

Examining the STEM inquiry activity that is selected for each specific condition gives us a glimpse of the resultant experiences of specific STEM lessons. For example, through Table 5.1, we can understand that this STEM lesson was designed with the intention for students to focus on real-life and authentic problems (e.g., the problem of increasing health care costs due to an increase in the number of chronic diabetic patients) and primarily apply the knowledge of science (understanding the chemical composition of diabetes medication, their half-life, and the mechanics behind how they are broken down by the body) and mathematics (mathematical modelling of variables such as diet, age, lifestyle, and family history affecting chances of suffering from diabetes) to make sense of the extent of the problem given. Students will then present their ideas (likely in the form of a predictive model as evidence with recommendations of changes in lifestyle) to users to gather feedback on the relevance of the proposed ideas.

From Chapters 6 to 9, we describe and interpret how students and teachers work together to enact integrated STEM inquiry from the conceptual lens of the STEM Quartet instructional framework and the four conditions for successful STEM inquiry. Each of the chapters will start with an introduction to the context, mapping of the STEM Quartet, a teaching vignette, and our interpretation of the interactions involved.

Questions for Reflection

1 What is the role of data and evidence in integrated STEM inquiry?
2 What are some variations that can be made using the three variants of the STEM Quartet instructional framework during lesson enactment?

Table 5.1 A hybrid application of the three variants of the STEM Quartet instructional framework (4Cs × Problem-Solution-User)

Conditions	Problem-centric STEM inquiry	Solution-centric STEM inquiry	User-centric STEM inquiry
Connections to real-world problems	• Make connections to contemporary issues within specific local contexts. • Problems are presented as complex, persistent, and extended.	• Consider interactions between the students and objects around them. • Present existing artefacts, technologies, or solutions to problems that are currently used in society.	• Interaction with members of the community to understand their experiences and diagnose the problems they face.
Opportunities for presentation and critique of ideas and claims	• Apply domain knowledge related to science and mathematics to state and critique claims. • Critique also involves understanding the context in which the problem is located.	• Apply domain knowledge related to engineering and technology to critique design and workability. • Knowledge is also used to understand affordances, strengths, and limitations of current designs/ solutions/ways of doing things.	• Apply knowledge of user needs and current social context(s) to make claims and critique an idea/a product/solution. • Compare existing policies and rules to the desired success criteria and user expectations.
Space for design, making, and investigation	• Students choose the materials they find most appropriate to design solutions. • Students test solutions by carrying out experiments and use the data to re-examine the problem and the solutions.	• Students work with existing prototypes/models to understand how they work and where changes/adaptations can be made. • Students test out their improved design against success criteria.	• Students consider ideas such as implementation details, accessibility, evaluation, and feedback mechanisms when users use the product/solution.
Modelling communities of STEM problem solvers	• Students work with experts (such as scientists or mathematicians) to learn techniques required to create solutions to the problems.	• Students work with industrial partners or suppliers to understand existing product features, intentions, and limitations.	• Students work with users in the community to understand different needs to make adaptations on the ground. • Emphasis is on impact authenticity.

References

Teo, T. W., Tan, A. L., Ong, Y. S., & Choy, B. H. (2021). Centricities of STEM curriculum frameworks: Variations of the STEM Quartet. *STEM Education, 1*(3), 141–156.

Veal, W. R. (2002). Content specific vignette as tools for research and teaching. *Electronical Journal of Science Education, 6*(4), 1–16.

6 Working as a multidisciplinary team

Planning a successful integrated STEM lesson to solve problems is not an easy task for one teacher. The involvement of multiple disciplines in the problem-solving process suggests that working in teaching teams comprised of teachers with different expert knowledge would be more productive. Each teacher brings along their expert disciplinary knowledge and skills to plan meaningful activities and select appropriate strategies to guide students to learn, select, retrieve, and apply relevant disciplinary knowledge to make connections between the problem and solutions generated. The vignette described in this chapter describes how Ms Fanny and her colleagues worked as an interdisciplinary team, with each member contributing their individual areas of expertise, to create rich and transdisciplinary learning experiences for their students. The team planned a series of lessons by weaving specific subject-matter knowledge and disciplinary skills into a problem-centric lesson on wastewater treatment. In the planning stage, the team ensured that the problem presented must be a problem related to the real world, offering students opportunities to work together so that they could practice the valued 21st-century competency of communication and collaboration. The targeted specific learning outcomes that the team of teachers included in their lesson plan were: (1) recognise the steps involved in problem solving, (2) perform water quality test, and (3) gather information and evaluate wastewater treatment. As described in Figure 6.1, the issue is presented to the students in the form of wastewater polluting the school environment and left untreated before disposal. The learning outcomes for science come from various topics and include concepts of turbidity, light, filtration, flowrate, and properties of materials. For mathematics, the learning outcomes are less extensive (hence the thinner lines of the box), with students learning the relationship between time and distance in the measurement of flowrate, learning the various formulas for geometry, and appreciating the role of mathematical and computational thinking in coding. For engineering learning outcomes, the teaching team had two intentions

DOI: 10.4324/9781003422501-6

80 Working as a multidisciplinary team

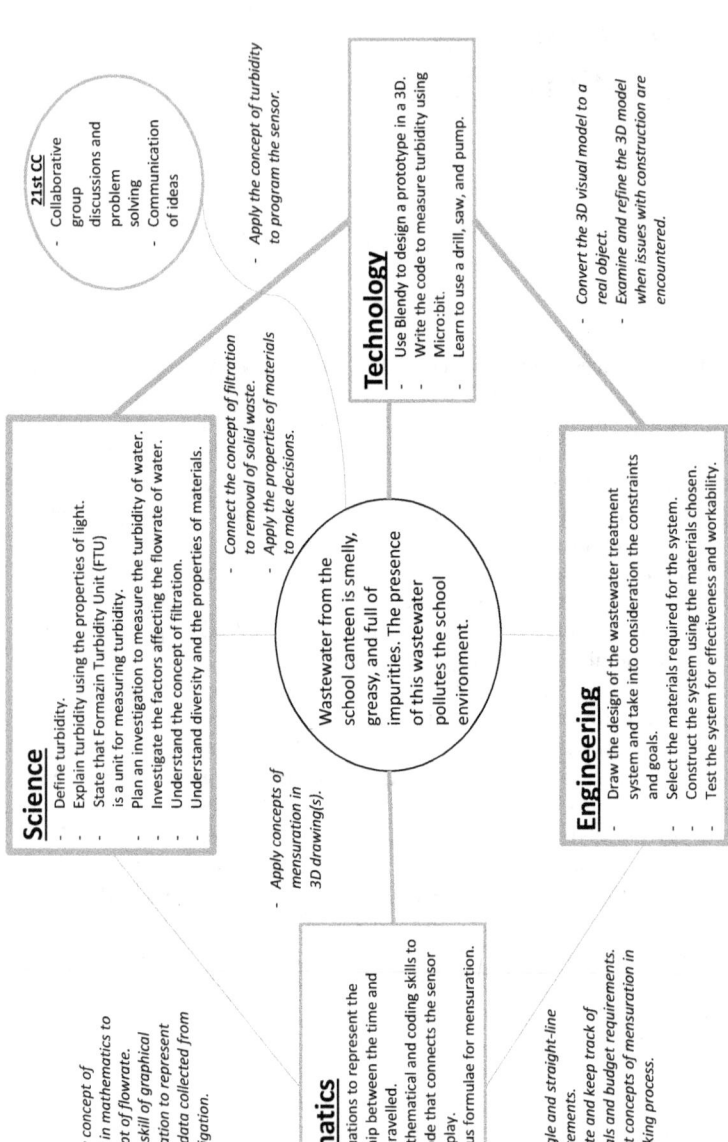

Figure 6.1 Representing wastewater treatment lesson using the STEM Quartet instructional framework.

for students to (1) experience designing a water flow and filtration system and (2) convert their design into a physical working prototype. Lastly, students were taught how to use the technological tool Blendy (Blendydomevj.com) to generate a 3D model and also how to do block coding using Micro:bit (microbit.org) to operate the turbidity sensor. The connection between science-technology and engineering-technology is stronger than that between mathematics-engineering and science-engineering for this activity. As the development of 21st-century competencies is highly valued in the school, the team also intentionally created spaces for students to practice collaborative and communication competency through this activity. The plan guided the team in their decision regarding expert responsibility, teaching strategies, time allocation, and resource acquisition.

In the enactment, besides acquisition of content knowledge such as turbidity, properties of light, filtration, and properties of materials, students also develop skills in designing, drilling, coding, and making. Opportunities were created to allow students to work collaboratively with others, help their peers, problem solve, and persevere when their plans failed. Throughout the process, students kept the success criteria in mind as they planned their prototype. Students spent time at home designing their prototypes. There were many opportunities for students to discuss in groups to improve their design and prototype. Through testing their prototypes, the students gathered data and used their observations to determine the areas that required improvements and made specific refinements to their design.

The instructional strategies used by the team of teachers include direct instruction, raising key questions, student-directed small group discussions, teacher-facilitated group discussions, hands-on building, online research, and investigations. Students' artefacts in the form of their prototype and presentations were used as evidence of students' learning.

6.1 Vignette 1—Working together as a teaching team

Interactions in the classrooms	Conditions observed
Vignette 1: Ms Fanny, Ms Pat, Mr Pong, Mr Gup, and Mr Trung's Story (pseudonyms are used)	
The Problem: Wastewater treatment system Ms Fanny and her team of teachers presented students with the task of designing and building a water treatment system to treat wastewater from the school canteen. The wastewater should be treated to reduce (1) odour, (2) suspended particles, (3) bacteria, and (4) grease. The class of 30 grade 8 students was allowed to use any materials they considered necessary as long as they spent within a budget of approximately USD190 for three cycles of iterations. To build students' skills to use tools such as electric drills and 3D rendering software, Blendy was taught to the students over two weekends. Students were given the details of the problem in class before they went home to do the necessary planning and research.	Relating to real-world context
STEM-inquiry processes As one of key success criteria of the treatment system is the clarity of the effluent (low concentration of suspended particles) collected from the filtration system, when students returned to class the next day, Ms Pat, the physics teacher, started the lesson by reviewing concepts related to light, scattering effects of light and introduced the word 'turbidity' to the students. She asked the class of 30 students who were seated in groups of five: 'What caused water to be turbid?' Students answered that turbidity could be due to suspended particles in water. Ms Pat acknowledged the students' answers and introduced the Formazin Turbidity Standards. After the introduction of turbidity and the standards for measuring turbidity, Ms Pat showed the students two bottles of water stored in plastic bottles. She asked the students: 'By looking at the water, which bottle of water do you think is safe to drink?' The students discussed this question in their groups and concluded that although the water looks clear, there could be dissolved substances and bacteria in the water.	Space for inquiry Claims and critique of ideas

Once the students were familiar with the concept of turbidity, Mr Pong, the computer teacher, handed turbidity sensors to the students. Mr Pong gave the students step-by-step instructions on how to assemble the different parts of the sensors and the rationale behind each part. The students were eager to explore and assemble the turbidity sensors as they were seen examining the sensors. They explored the various connections between the sensors and the pocket-size computer with their peers, raising their hands occasionally to ask Mr Pong to check on their connections.	Space for design, making, and inquiry
When the sensors were connected, the students proceeded to the computer laboratory, where they worked in pairs to alter a source code aimed at measuring the turbidity of water. They tested their codes (see Figure 6.2) and calibrated their turbidity sensors using water (very cloudy, cloudy, and clear) that the teachers collected earlier in the day.	Modelling communities of problem solvers

Figure 6.2 An example of the source code.

Following the testing of the sensors, Mr Gup, another computer teacher, introduced the students to the concept of flowrate, which will be used to evaluate students' design of their wastewater system. After a brief lecture on flowrate (see Figure 6.3), students worked in pairs to examine the flowrates of the various sections of the 3D design of their water treatment systems using Blendy. While students were working in pairs, they were seen questioning each other and generating alternatives to improve their designs. Students were frequently seen pointing at specific parts of their design on the screen and discussing. The lesson for the day ended with the teachers informing the students that they would be using their 3D design to build and test their solutions the next day.	Space for inquiry Modelling communities of problem solvers

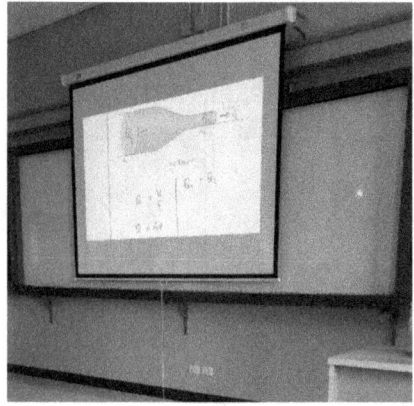

Figure 6.3 Understanding flowrates.

Making and Tinkering

The next day, the students entered the classroom with materials that they brought from home. Some groups brought used containers that previously stored paints, while others had purchased enzymes that digest fats. The lesson started with Mr Trung, the design and technology teacher, explaining the use of tools such as the drill, sticky tape, glue, and pen knives (see Figure 6.4). Safety precautions were also emphasised during this briefing.	Space for design, making, and inquiry

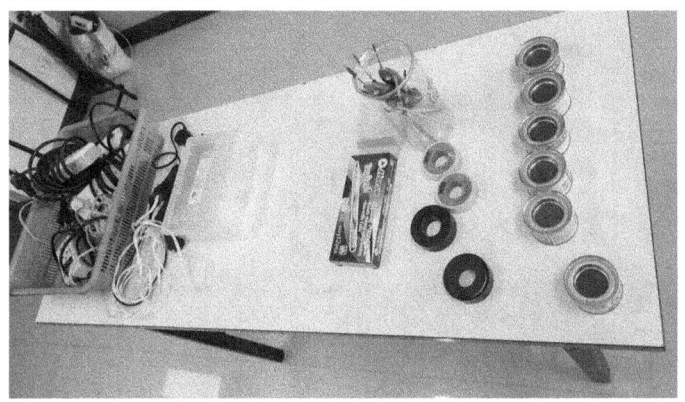

Figure 6.4 One of the three tableful of tools.

After the briefing, the students got to work in their groups. One group of students started drilling 16 evenly spaced holes on a flat surface to allow water to flow through. At the drilling station, students from different groups gathered together to help one another with the drilling process by stabilising the tables and the handles of the drill. While the students were competing with one another to design the best wastewater treatment system, they nevertheless were willing to help one another, and this collaboration was observed frequently.

The students did careful planning of their work before the lesson. One group of students brought in a mounting board with the different parts drawn to scale on only one part of the mounting board (see Figure 6.5). They asked another group of students to share half the mounting board with them so that both groups could keep within the allocated budget. This form of sharing was common among the students.

Modelling communities of problem solvers

86 *Working as a multidisciplinary team*

Figure 6.5 Mounting board that was shared.

At the back of the classroom, one group of students found that the containers that they purchased were too small and low and as such, they were unable to create a gradient for water to flow downwards. The students tried to solve the problem by scaling down their treatment system but failed to create the necessary gradient and force for water flow. After three attempts, they decided to consult Mr Pong to explore other possible solutions (see Figure 6.6). Eventually, the group had to look for larger containers within the school compound to build their prototype.

Modelling communities of problem solvers

Working as a multidisciplinary team 87

Figure 6.6 Troubleshooting with students.

After an intensive two hours of work, the students left their workstations in the laboratory and proceeded to the assembly area of the school to test their prototype. The testing process was systematically carried out. First, the students check for leaks by running water through their wastewater treatment system (see Figure 6.7). One group of students realised that water leaked out at every joint area! They used tape to reinforce their joints and checked for leaks again. Once all the leaks were patched, the groups checked the flowrates by timing the time it takes for water to move from the first tank until it emerges.	Space for inquiry

88 *Working as a multidisciplinary team*

Figure 6.7 Checking for leaks.

Once the prototypes were tested and obvious faults fixed, the students presented their wastewater treatment system to their peers. They explained their various parts of the design and the rationale for including each part. The peers raised questions and offered suggestions for improvements.	Critique of ideas

6.2 Relating vignette to conditions for successful integrated STEM inquiry

Water treatment is a real-world situation since availability of clean and potable water is needed in many situations. Clean water and sanitation are the focus of sustainable development goal 6 in the United Nations Sustainable Development Goals (UNSDG). One of the key challenges that teachers face is to identify a real-world problem that relates to students' experiences. One effective way of doing this is to start with the students' immediate environment such as the school, the estate they live in, or places that the students frequent. Intentionally connecting their daily and familiar experiences to the real-world context helps students appreciate the wider implications of their individual actions. The problem identified in vignette 1 is located in the school canteen. The school canteen is a familiar place for the students as they visit the place daily for their meals. While the students know of the presence of wastewater, they do not have in-depth knowledge about the fate of the wastewater generated in the canteen. By introducing the context of wastewater in the school canteen, the teachers engaged the students with familiar experiences in their daily lives to begin the integrated STEM inquiry (taking condition 1 into consideration). This lesson adopted the problem-centric variant as the students were presented with the problem of wastewater in the canteen. They were not given any specific model of water treatment at the start of the activity.

To equip students with the necessary domain-specific knowledge (behaviour of water, turbidity, principles of filtration, and investigation of flowrates) to construct and critique claims related to water treatment and water quality, whole class instructions, group discussions, and teacher-directed lectures were used. For instance, the physics teacher, Ms Pat lectured on the concept of turbidity and defined it as the relative clarity of a liquid. These ideas of turbidity were also related to ways of measurement of turbidity in the real world. In this instance, the domain-specific knowledge relates to physical science ideas of light scattering and light detection. Students learn about the equipment used to detect the amount of light scattered. The measurement of turbidity using a light sensor positioned in its direct path is an example of how domain-specific knowledge is connected to domain-specific skills of science inquiry. Students' experience with the turbidity sensor as a tool (material aspect) to detect the path and intensity of light is one example of how the epistemic practice of the empirical nature of science inquiry can be orchestrated. The learning of domain-specific knowledge was also interspersed with students clarifying the success criteria and understanding constraints. These instances demonstrated students' engagement with the epistemic aspect of pedagogical practice in integrated STEM inquiry.

Further, in the vignette, the computer and technology teachers worked with the students to practice skills such as drilling holes to allow water to pass through. Students learn to choose drill bits with appropriate sizes, attach the drill bits to the drill, and devise ways to stabilise the object when the drill vibrates. Learning and perfecting these skills closely related to engineering making processes enable students to build high-quality prototypes of wastewater treatment that were subsequently tested for workability. These domain-specific skills and practices in engineering complement the scientific knowledge related to turbidity and behaviour of light to facilitate students' sense-making of the problem. The making part of the lesson is an example of the material aspect of successful STEM inquiry and also an example of the pedagogical practice of pacing and maintaining momentum of an iterative problem-solving cycle, where the conceptual and procedural aspects of learning needful disciplinary-based knowledge and skills are taught.

Observing the lesson, it was also evident that Ms Fanny and her team designed the learning experiences such that they afford opportunities for students to work in competitive and collaborative teams. The competitive nature of the activity is created by the success criteria set for the activity. Groups will be judged, and the best wastewater treatment system will be recognised. Each group is given a fixed amount of budget to build their prototype, and the most prudent group will be rewarded eventually. To cut costs, the students began to collaborate across teams to co-pay and share items that they required. This collaborative behaviour was prevalent and propelled students to move forward as a class. This social aspect of working in teams towards a common goal was very clearly demonstrated by students.

Finally, Ms Fanny and her team used various forms of evaluation to assess students' progress and learning. As part of the inquiry process, each prototype was subjected to test as students sought to obtain data on the flowrate of their system to evaluate its efficacy. Using students-led investigations prompted students to make improvements to their design based on criteria and observations chosen by themselves. While each team made decisions on how they would like to conduct their testing, many teams chose similar variables to test. Feedback was also given by the team of teachers who served as the knowledgeable others in this problem-solving community. The teachers raise questions related to specific aspects of students' design to encourage students to think deeper about the finer parts of their design. The feedback and evaluation serve to engage students in reflection and make salient the conceptual/procedural, epistemic, and social aspects of their STEM learning experience.

Table 6.1 shows the variation in enactment of the problem-centric lesson on solving the problem of wastewater treatment. The problem is presented as a real-world problem that is complex, persistent, and extended and is located within a context that is easily accessible by the students. There were many opportunities created for students to discuss and critique in order to learn deep domain-specific knowledge related to science, coding, engineering, and mathematics. As such, the opportunities presented for presentation and critique have characteristics of both problem- and solution-centric variants. The testing of the wastewater prototype resembles of problem-centric STEM inquiry since the students tested their prototype by carrying out investigations before they made refinements. Students were able to freely interact and be guided by a team of teachers with expertise in programming, design, technology, mathematics, and science. This community of problem solvers is characteristic of a problem-centric STEM inquiry variant. In this activity, there was no evidence that students were engaged in consideration of users.

Table 6.1 Understanding the variation in practice for vignette 1

Conditions	Problem-centric STEM inquiry	Solution-centric STEM inquiry	User-centric STEM inquiry
Connections to real-world problems	• Make connections to contemporary issues within specific local contexts. • Problems are presented as complex, persistent, and extended problems.	• Consider interactions between students and objects around them. • Present existing artefacts, technologies, or solutions to problems that are currently used in society.	• Interact with members of the community to understand their experiences and diagnose the problems they face.
Opportunities for presentation and critique of ideas and claims	• Apply domain knowledge related to science and mathematics to state and critique claims. • Critique also involved understanding the context in which the problem is located.	• Apply domain knowledge related to engineering and technology to critique design and workability. • Knowledge is also used to understand affordances, strengths, and limitations of current designs/solutions/ways of doing things.	• Apply knowledge of user needs and current social context to make claims and critique. • Use existing policies and rules to compare with the desired success criteria and user expectations.
Space for design, making, and investigation	• Students choose the materials they find most appropriate to design solutions. • Test solutions by carrying out experiments and using the data to re-examine the problem and solutions.	• Students work with existing prototypes/models to understand how they work and where changes/adaptations can be made. • Test out improved design to meet success criteria.	• Students consider ideas such as implementation details, accessibility, evaluation, and feedback mechanisms when users use product/solution.
Modelling communities of STEM problem solvers	• Work with experts (such as scientists or mathematicians) to learn techniques required to create solutions to the problems.	• Work with industrial partners or suppliers to understand existing product features, intentions, and limitations.	• Work with users in the community to understand different needs to make adaptations on the ground. Emphasis is on impact authenticity.

7 Learning from nature's design

In solution-centric STEM inquiry, teachers present students with existing solutions to current problems, and students evaluate the solutions to distil insights about the affordances, advantages, and disadvantages of the solutions. Solution-centric STEM inquiry works on the assumption that there are no perfect solutions to problems. Rather, every solution has limitations, and there are trade-offs made. As such, uncovering these limitations could allow learners to learn how current solutions work and how they are created, and explore other alternatives. In vignette 2, we describe a lesson that shows characteristics of solution-centric STEM inquiry with the teachers obtaining their inspiration from nature's design. The aim of the series of lessons was to introduce students to the concept of biomimicry and apply the principles to engineering designs. The biomimicry approach can be found in applications such as Velcro (mimicking seeds with hooks), industrial adhesives (molluscs), and turbine blades (whale fins).

Using the STEM quartet solution-centric variant diagram showed a strong emphasis and connections on science and technology learning outcomes. There are comparatively weaker learning outcomes for mathematics and technology and hence the connections to these epistemic practices are also correspondingly weaker. Figure 7.1a and 7.1b depict the plan for the lessons on design from nature.

Learning from nature's design 93

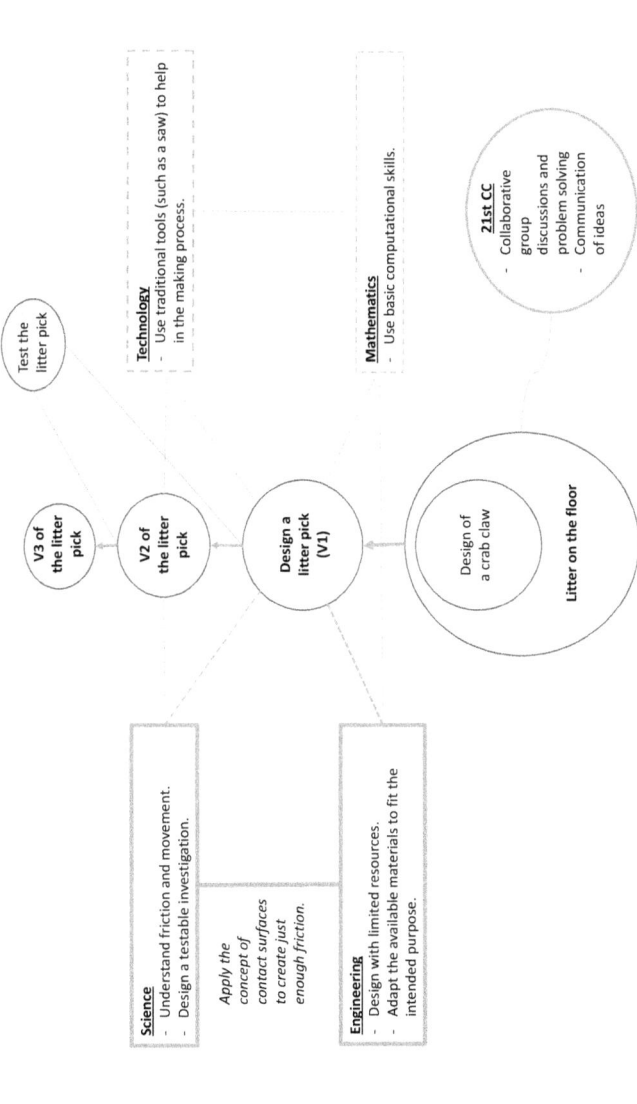

Figure 7.1 (a) Representing biomimicry design of litter pick.

94 *Learning from nature's design*

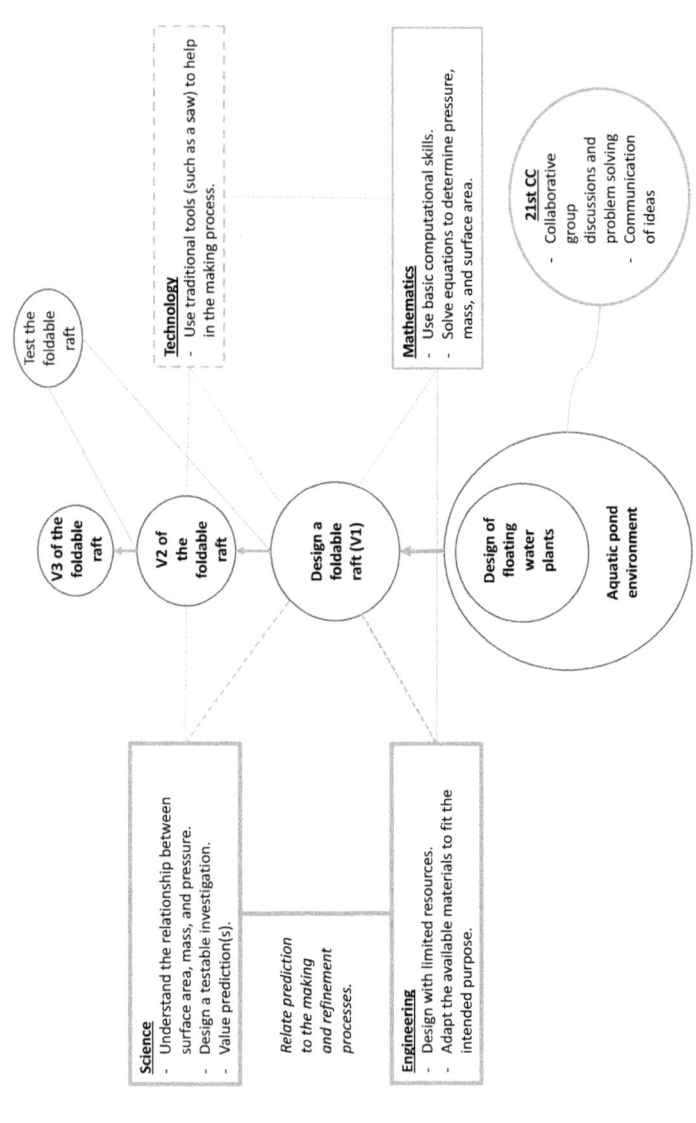

Figure 7.1 (b) Representing biomimicry design of foldable raft.

Note. V1, V2, V3 represent Version 1, 2, 3, respectively.

7.1 Vignette 2—Biomimicry

Interactions in the classrooms *Vignette 2: Mr Jet and Mr Pu harnessing design from nature*	Conditions observed
The litter-picking challenge! Mr Jet began the lesson by posing a litter-picking challenge to his students. He gave students a container with a fixed quantity of typical litter, which was first poured out onto a flat surface, and then students had a fixed time to pick up as many pieces as possible. In order to accomplish this challenge, the students were first introduced to the principle of biomimicry through a lecture and some activities, and the general method of design problem solving. Students were organised into teams of three students each, and the whole group was assigned some brief research activities in which they had to present their work to their peers. Students were generally motivated and inspired, cutting short their lunch hour to return to work on their projects. The lesson was carried out in the school hall, which had a good sound system; teachers Mr Jet and Mr Pu were on the microphones reminding students of their tasks and giving helpful advice on how to proceed. Overlaid onto a cheerful music soundtrack, the overall effect was to give the room a carnival-like atmosphere, which lowered the tension in the room and created the conditions more amenable for the kinds of divergent thinking necessary for creative activities. With students sprawled all over the floor working on their presentations, the lively hall atmosphere certainly contrasted against the sombre, unoccupied tenor of the rest of the school. The school instituted a 'materials budget' for students' projects: instead of a 'free flow buffet' of tools and materials, they were assigned a certain number of credits that they could use. This method dramatically reduced material usage and the artefacts produced showed more efficient and mindful use of materials. Besides this, the resource limitation made it such that students were then committed to communicating to one another why their design proposal was superior and why certain materials or handling techniques (tool use) were preferable to others. To do so, a lot more sketching was done, and these sketches (as representations of ideas) were instrumental in helping students become clearer about what their project was.	Relating to real world Presentation and critique of ideas

Once the students completed their design and sketching, they proceeded to make their prototype. The entire hall essentially became transformed into an elementary makerspace, with the tablecloths and skirting partially removed so that pieces of wood could be worked on with saws, cardboard could be sliced with box cutters, and hot glue stations to help students put everything together. The students started using the coping saw provided, a type of saw with a very narrow (about 2 mm width), and actually quite brittle blade. Within the first few minutes of use, it quickly became apparent that they lacked the skills to handle the saw. Several groups almost simultaneously cried out that their blades had broken. Here, in a very direct manner was their experience of the mangle of practice, in which their expectations of the way in which their apparatus was to work contrasted dramatically against the material reality of the things in front of them. Also, barely visible (but certainly discernible by touch) was the sawtooth pattern of the blade: while some attempted to replace the blades by themselves, most approached the teacher aide for assistance. It took several tries (and broken blades) for students to realise the direction the saw was cutting, and that they should exert more force, versus the other direction, in which less force was to be exerted and more control was needed to prevent breakage. After an afternoon of making, 10 of the 20 groups of students produced variations of the pincer design, which was a well-established design. Some groups produced designs with notable exceptions such as scissor mechanisms (which tended to fail because the two claws failed to meet at a common point) and one interesting mechanism which emulated musculature with flexor (muscle that flexes a joint) and extensor (a muscle whose contraction extends or straightens a limb or other parts of the body) components, the ingenious use of drinking straws as a spring element, and sandpaper-covered claws to increase grip (See Figure 7.2).	Space for design, making and inquiry Space for inquiry

Learning from nature's design 97

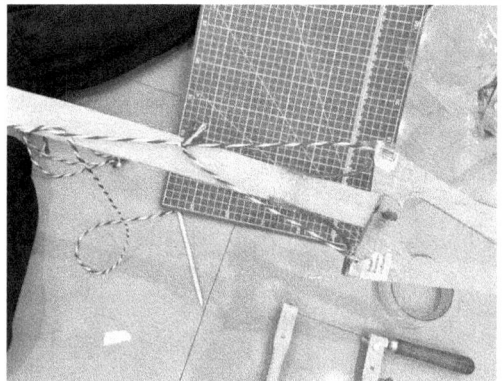

Figure 7.2 An example of a prototype to pick litter.

One team declared that their design was inspired by crab claws (see Figure 7.3), and although it seemed to work well as a design, the students did not explain what functional aspects of crab claws inspired this design. Student designs were tested against the standard scenario of picking up litter; through this process, flaws in their design instantly became obvious. The testing process highlighted the gap between their intentions and aspects of nature, which they had been ignorant of. Mr Jet and Mr Pu informed the students to return to their groups to make revisions to their projects (admittedly minor), before the end of the day. The next morning began with the final testing of the student projects and declaring the winner of litter-picking challenge.

Critique of ideas

Figure 7.3 A crab claw prototype.

98 *Learning from nature's design*

Foldable Raft Challenge!	
After the experience with designing the litter picker, the students were introduced to the second task of creating a foldable raft. For the foldable raft challenge, Mr Jet and Mr Pu prepared blocks of plasticine of known mass as prototypes were tested for how many pieces they could hold up before a piece of tissue paper placed on the deck of the raft was observed to become wet. Similar to the litter pick challenge, there was a brainstorming session, an opportunity to study the real-life samples (freshwater crab, lotus leaves, water hyacinth, and others), followed by a presentation to other teams in their breakout group (15 students per group). Then there was a time for their plans to be revised, before 'purchasing' their items, and then students proceeded with making their prototypes.	Relating to real world Space for inquiry
In this second challenge, students understood better the imperative to "fail often so that you can succeed earlier", and quickly tested their minimally viable prototypes in the test apparatus provided—a large basin filled with water. Fairly quickly, obvious mistakes could be weeded out, such as the hinge element located at the upper, deck surface of the raft. This led students to apply reinforcing ribs across the hinge. During the testing period, it was interesting to watch as students almost universally underestimated the number of blocks that their prototype could support. For instance, Figure 7.4 shows the group that predicted that their raft could only support two blocks.	Critique of ideas
 *Figure 7.4* Testing prototype.	

Instead, this raft turned out to be one of the best performing, supporting up to 16 blocks before the piece of tissue paper (as indicator) got wet. A score could be calculated as the number of blocks less the absolute value of the difference between the number and the prediction: Score = number of blocks − ABS (number − prediction).	

7.2 Relating vignette to conditions for successful integrated STEM inquiry

Vignette 2 is a peculiar way to approach STEM inquiry from a solution-centric perspective as it focuses on design, not on novel design but on design 'borrowed' from nature. In vignette 2, the teachers engaged students' interests and experiences by choosing a familiar real-world activity of picking up litter from the school compound and for the second activity, to design a foldable raft. Litter is everywhere in schools and along the sides of streets, while a raft is something that students would have been exposed to during their childhood games. Beginning with these two familiar scenarios and referencing equally common objects such as crab claws, lotus leaves, and water hyacinths greatly increased students' connection of the challenge. This also describes the pedagogical practice related to introducing the context to students through introducing the concept of biomimicry.

A large proportion of class time was allocated for students to construct the two artefacts. Hence, the material aspects of learning were evident—students learn how to use drills, saw, work, and adapt materials available to them and convert 2D design on paper into 3D artefacts. As making and testing are tightly coupled, students also tested their designs often. The presence of concrete 'litter' and the basin of water for testing of floating devices offered plenty of opportunities for students to obtain data for discussion and critique ideas. This testing forms part of the inquiry process to gather data that is fed forward into the refinement process.

During the lessons, there were many opportunities for students to engage with peer critique to learn the domain-specific concept of biomimicry. For instance, students' lack of clear understanding of biomimicry at the start might have led to the possible interpretation of biomimicry as shape borrowing instead of function borrowing. In the former, a common approach in Design and Technology classrooms, bio-inspired (or other) designs are mostly applied as a gloss over an unrelated mechanism. However, in function borrowing, the artefact is essentially inoperable without the inspired design. Admittedly, this distinction can sometimes be hard to

100 *Learning from nature's design*

discern, but in several of the cases of students developing pincer variants, it seemed that the biomimicry tended to be more of the shape borrowing rather than the function borrowing sort. The need for introduction of conceptual/procedural knowledge in presenting the activity is evident from this lesson. Unless the students understood the concept and principles of biomimicry, they would not be able to proceed with improving the design in a meaningful manner.

Similar to Ms Fanny and her team in vignette 1, Mr Jet and Mr Pu provided students with a fixed number of credits for making their prototype. Hence, students were given an opportunity to develop their competencies in budget planning, negotiations, and decision making while designing a litter picker that resembles design in nature. There was little evidence of a systematic way of evaluating students' learning in this lesson. When students were testing their prototype, there was no systematic way of assessing the success criteria. It was only upon discussion that the teachers realised the need for a scoring system to be developed and hence they devised a way to incorporate the difference between students' predictions and test performance as a means to measure how well student designs had been tested. The development of a scoring system is an example of the epistemic aspect of pacing and maintaining momentum of an iterative problem-solving cycle since it enables students to revisit their designs and improve on their ideas. There is also a social aspect to this when students are put in a position where they have to be responsive to feedback given by their peers.

In considering each condition of successful integrated STEM inquiry, the lesson described in vignette 2 started with observations of nature's design around the students. This is a characteristic starting point for solution-centric STEM inquiry. Students were presented with ready artefacts, which is a feature of solution-centric STEM inquiry. The domain knowledge focused largely on the design in nature and how these designs can be replicated to create better artefacts and the students tested the artefacts that they produced. What is clearly lacking in this vignette are the opportunities for modelling community of STEM problem solvers or designers. Students could watch a video of how inventors make observations of nature to inspire their design. Vignette 2 shows three of four characteristics of a typical solution-centric inquiry (see Table 7.1).

Table 7.1 Understanding the variation in practice for vignette 2

Conditions	Problem-centric STEM inquiry	Solution-centric STEM inquiry	User-centric STEM inquiry
Connections to real-world problems	• Make connections to contemporary issues within specific local contexts. • Problems are presented as complex, persistent, and extended problems.	• Consider interactions between students and objects around them. • Present existing artefacts, technologies, or solutions to problems that are currently used in society.	• Interact with members of the community to understand their experiences and diagnose the problems they face.
Opportunities for presentation and critique of ideas and claims	• Apply domain knowledge related to science and mathematics to state and critique claims. • Critique also involved understanding the context in which the problem is located.	• Apply domain knowledge related to engineering and technology to critique design and workability. Knowledge is also used to understand affordances, strengths, and limitations of current designs/solutions/ways of doing things.	• Apply knowledge of user needs and current social context to make claims and critique. • Use existing policies and rules to compare with the desired success criteria and user expectations.
Space for design, making, and investigation	• Students choose the materials they find most appropriate to design solutions. • Test solutions by carrying out experiments and using the data to re-examine the problem and solutions.	• Students work with existing prototypes/models to understand how they work and where changes/adaptations can be made. • Test out improved design to meet success criteria.	• Students consider ideas such as implementation details, accessibility, evaluation, and feedback mechanisms when users use product/solution.
Modelling communities of STEM problem solvers (None)	• Work with experts (such as scientists or mathematicians) to learn techniques required to create solutions to the problems.	• Work with industrial partners or suppliers to understand existing product features, intentions, and limitations.	• Work with users in the community to understand different needs to make adaptations on the ground. • Emphasis is on impact authenticity.

8 Growing humanistic values through integrated STEM inquiry

In vignette 3, we illustrate how a team of teachers worked together using their areas of expertise to create a rich integrated STEM inquiry learning experience to engage students' hearts, minds, and hands. It was noteworthy that while the lesson's intention and instruction were focused on the development of specific subject-matter knowledge and disciplinary skills, these knowledge and skills were contextualised in a real-world context using a problem related to mountain pipelines. You will observe that the flow of the lesson from learning domain-specific knowledge to problem-solving was seamless, and students had opportunities to relate what they had learned to the problem at hand immediately. The enactment of the lesson offered many opportunities for students to develop their skills in critical thinking, communication, and collaboration. More importantly, the students gained a deeper appreciation for the complexity of making water available in mountainous regions. Accessible water supply is usually taken for granted by many and, hence, few would be able to empathise with the problems or difficulties faced by villagers living in mountainous regions. Contextualising learning with 'familiar yet detached' contexts offers students opportunities to engage with issues that they are less likely to encounter by themselves or may not be fully aware of. Through learning about the issues, students grow their knowledge and develop greater sensitivities on how knowledge and practices from different disciplines can serve as alternative lenses for problems. The learning experiences also result in students gaining an enriched perspective of the problem which could potentially serve as a stimulus for innovative solutions to other wicked problems of the 21st century. Through this activity, students were able to understand the humanistic perspective of integrated STEM inquiry—STEM learning is not just a school subject but can have implications beyond gaining subject-matter knowledge as it can be used to improve the lives of communities.

The plan for mapping the activities on user-centric STEM inquiry is seen in Figure 8.1.

Growing humanistic values through integrated STEM inquiry 103

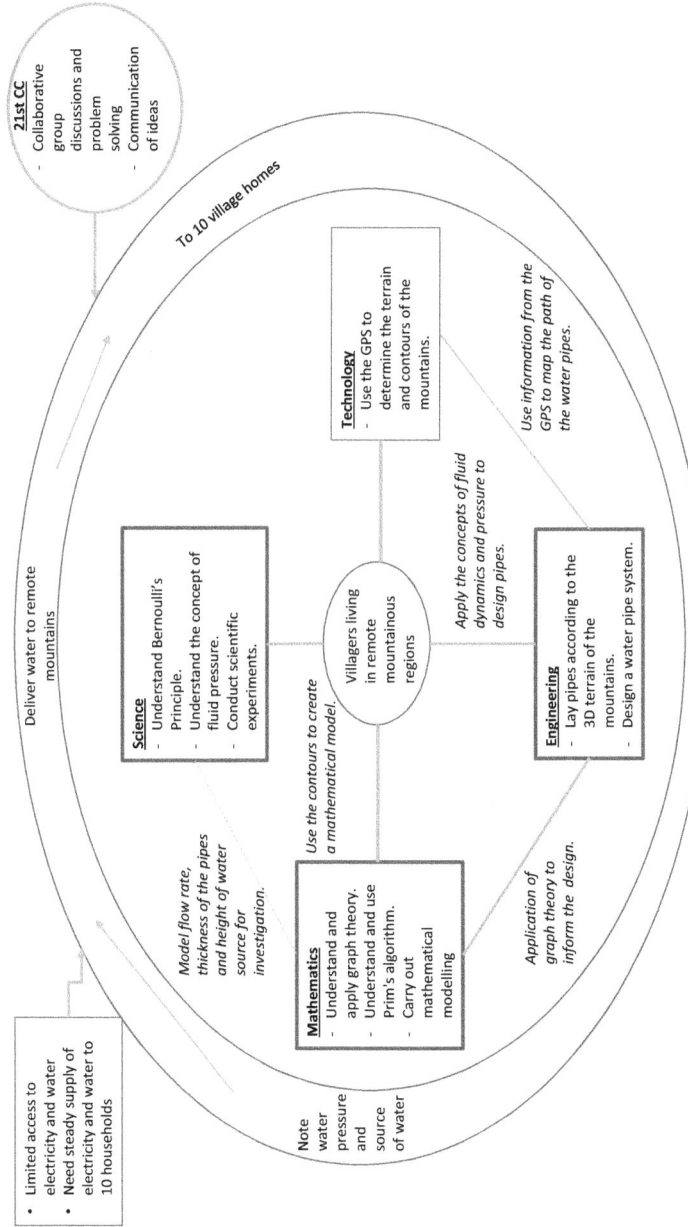

Figure 8.1 Mapping learning outcomes of user-centric STEM.

8.1 Vignette 3—Humanistic STEM

Interactions in the classrooms *Vignette 3: Engaging hearts, minds, and hands in integrated STEM inquiry*	*Conditions observed*
Engaging the hearts in integrated STEM inquiry	
Mr Teerute and his team of teachers developed a user-centric integrated STEM unit focusing on issues related to delivering water to villagers located in remote mountainous regions. Villagers living in remote areas have limited access to electricity and water, and Mr Teerute and his colleagues decided that it was a meaningful context for students to learn about Bernoulli's Principle, Graph Theory, and Map Reading by working with the needs of villagers as they develop a mountain pipeline that will supply water to ten households living in one of these villages.	Relating to real world
The context was presented by Ms Ketsarin using snippets of local news broadcasts to support students in empathising with issues and challenges faced by the villagers. The villagers featured in the clips raised students' awareness of the challenges of living in mountainous regions from the villagers' perspectives (See Figure 8.2). 	Relating to real world
Figure 8.2 Using videos to raise students' awareness.	

The second video highlighted the issues of water supply in Thailand to drive home the important message of providing clean portable water to everyone. The last video clip provided insights into how one village managed their encounters with water supply in the mountains and students realised the challenges of designing water pipe systems in the mountains. The students became aware of the importance of considering water pressure and the piping layout. The villager's solution also highlighted some issues with the existing piping solution, which suggests the need to look into these problems more carefully. Overall, these videos provided the stimuli for group discussion via graphic organisers (Figure 8.3) and whole-class discussions (See Figure 8.4) through the 'thinking board' (See Figure 8.5). The discussion of these issues not only helps students to see the 'humanistic' aspect of the water supply problem but also sets the stage for the topics to be introduced later.	Relating to real world
 *Figure 8.3* Group discussion using graphic organiser.	Critique of ideas

106 *Growing humanistic values through integrated STEM inquiry*

Figure 8.4 Whole-class discussion.

Figure 8.5 Students sharing their ideas documented on their thinking board.

The empathising part of the lesson naturally introduced the first issue of how the pipes should be laid out—the application of graph theory to solve real-world problems. Mr Blue recapped some basic concepts such as graphs, nodes, weighted graphs, and minimal spanning trees and introduced one of the greedy algorithms (Prim's algorithm) to students via an inquiry-based mathematics task. The task involved two related problems for students to investigate and work out the solution. These two tasks (See Figure 8.6) prepared the students to think about the application of graph theory to solve 'Problem of Mountain Water Pipeline System for Faraway Village' (See Figure 8.7).	Relating to real world

Growing humanistic values through integrated STEM inquiry 107

> Group

Activity Sheet 4.2
Telephone Transmission Lines Problem

A telephone transmission lines company want to connect a telephone line the villages A, B, C, D, E and F. They plan to run a telephone line along the road. If the cost of running a telephone line depends on the length of the telephone line, how will this company run a phone line at the lowest cost? By given a table showing the distance (km) of the road connecting the villages.

Villages	A	B	C	D	E	F
A	-	30	-	-	-	40
B	30	-	10	-	50	20
C	-	10	-	20	30	-
D	-	-	20	-	10	20
E	-	50	30	10	-	60
F	40	20	-	20	60	-

1) Students use the given information to draw a map by using the geometer's sketchpad.
2) Students use the map they drew to solve problems.
3) Students record their ideas and do the activity worksheet.

Activity Sheet 4.3
The Shortest Flight Path

How do pilots find a route from City A to City J with the least amount of flight time? By given the time it takes to fly (hours) from one city to another.

	A	B	C	D	E	F	G	H	I	J
A	0	7	0	0	0	0	8	5	0	0
B	7	0	8	0	0	0	0	4	3	0
C	0	8	0	8	0	0	0	0	5	0
D	0	0	8	0	3	0	0	0	0	4
E	0	0	0	3	0	1	0	0	7	0
F	0	0	0	0	1	0	7	0	0	6
G	8	0	0	0	0	7	0	5	2	0
H	5	4	0	0	0	0	5	0	0	0
I	0	3	5	0	7	0	2	0	0	0
J	0	0	0	4	0	6	0	0	0	0

1) Students use the given information to draw a map by using the geometer's sketchpad.
2) Students use the map they drew to solve problems.
3) Students record their ideas and do the activity worksheet.

Figure 8.6 Learning tasks for graph theory.

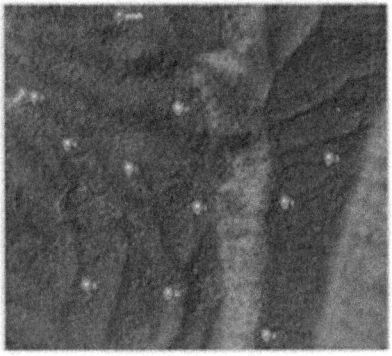

Figure 8.7 Challenge problem for the STEM unit.

Engaging the minds in integrated STEM inquiry

The students were highly engaged with the two mathematics learning tasks as they made sense of the information to see how their knowledge of graph theory could be applied to solve the two mathematics-based problems and, ultimately, the development of the water pipeline. They discussed in groups and collaborated on representing the scenarios using tree diagrams. Mr Blue then invited the students to share their answers to the learning tasks. The students were given time to discuss the main issue related to the design of the water pipeline.	Critique of ideas

Growing humanistic values through integrated STEM inquiry 109

As can be seen from Figure 8.7, there was little information shared with the students on the activity. Instead, the link to the actual Google Maps was given. Hence, the students had to make sense of the map and model the location of the ten houses using graph theory. These learning experiences provided opportunities for students to engage with the mathematical modelling process—an important process of thinking with mathematics. They measured distances on the map using the in-built functions within Google Maps and modelled the situation using graph before they realised the problem was analogous to the ones given in the learning tasks. The students also engaged with mathematical communication and representation to help others make sense of the problem (See Figure 8.8) before they presented their initial graph theory solution to determine the 'optimal' pipeline connections.	Modelling communities of problem solvers
 *Figure 8.8* Students modelling the problem using graph theory.	Critique of ideas

110 *Growing humanistic values through integrated STEM inquiry*

Having determined the 'supposedly' optimal pipeline connection, it quickly dawned upon the students that they had to consider how the water could be transported through the pipelines. To that end, Mr Teerute introduced the concept of fluid pressure to the class. The introduction of this problem was given in the form of a speed challenge that involved students 'transferring' water from a water bottle to a beaker with only a straw in the shortest time possible! This speed challenge provided students with an embodied experience of the concepts of fluid dynamics, which would be useful for designing the pipeline system. The students were engaged, and they came up with different ways to do so before Mr Teerute showed how pressure could play a critical role in 'transferring water' (see Figure 8.9).	Critique of ideas Space for design, making, and inquiry

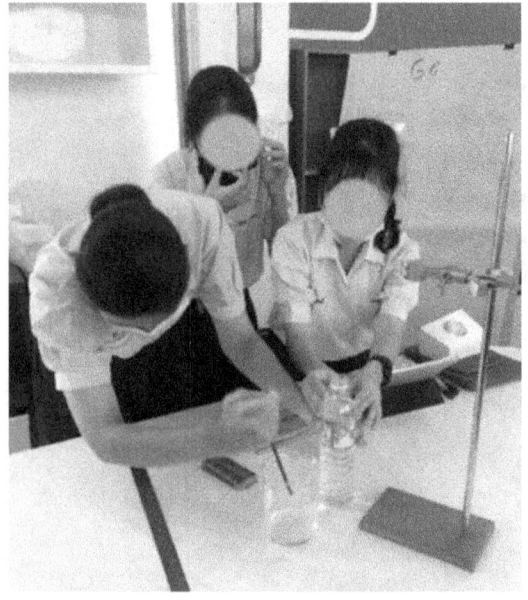

Figure 8.9 Students working on the 'water transfer speed challenge'.

Growing humanistic values through integrated STEM inquiry 111

Mr Teerute introduced Bernoulli's principle to the students through a series of guided experiments and discussion. The students were split into different groups to work on different experiments that focused on a specific set of conditions. The idea is similar to the jigsaw method that required different groups of students to share their findings with the whole class to determine the relationships between flowrate, thickness of pipes, and height of water source (see Figure 8.10). The students were encouraged to think about how their experiments and the findings shared were related to the problem of the water pipelines.

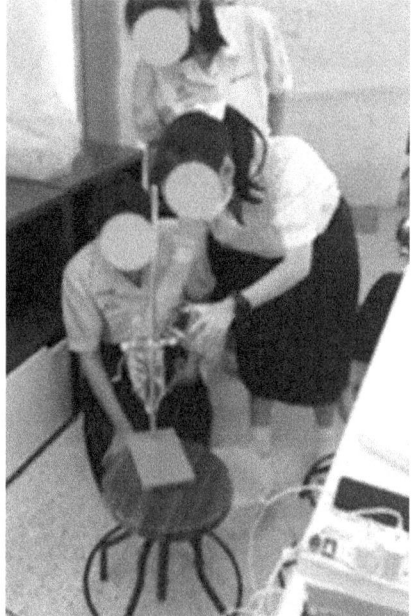

Figure 8.10 Students working on the experiments.

Of particular interest is how Mr Teerute referred the students back to one of the video clips that was used by Ms Ketsarin to highlight the issues with an existing solution deployed by one of the villages. In that village, the people complained about low water pressure when the water reached homes. Mr Teerute then explained how Bernoulli's principle could be used to account for what happened in the village (See Figure 8.11).

Space for inquiry

112 *Growing humanistic values through integrated STEM inquiry*

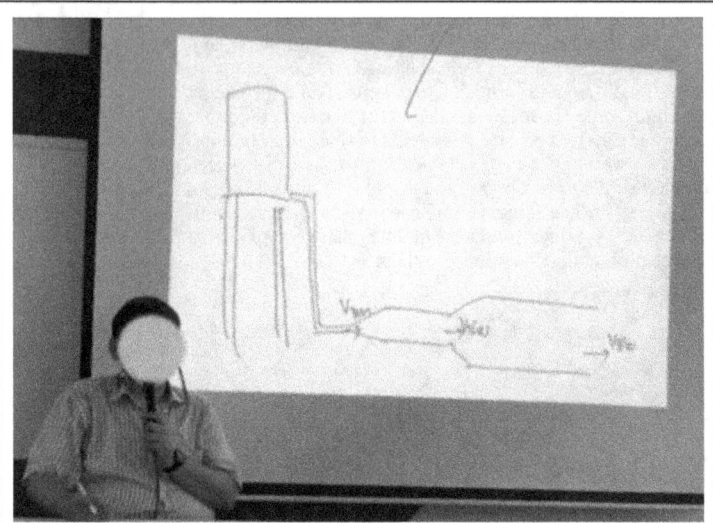

Figure 8.11 Mr Teerute relating the Bernoulli principle to the village's solution.

The in-depth domain knowledge of Bernoulli's principles prepared the ground for students to realise the need to consider the contour and terrain of the mountains in which the houses were located. A series of key questions were asked by Mr View for students to think about the different considerations needed to build a water pipeline system. Furthermore, in order for the students to think more deeply about the complexity of the problem, they were asked to build a scaled model of the actual mountain (See Figure 8.12). The students had to use the information embedded in Google Maps and the contour lines to build a scaled model of the mountains.

Growing humanistic values through integrated STEM inquiry 113

Let's make the Clay Mountain!!!

From the given contour line Map, your group is required to make the mountain that have the height and width in the same ratio of the actual size.

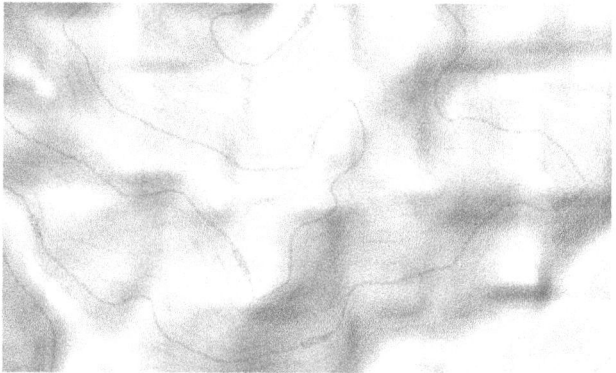

Contour Map show geological structure around Na Halo District, Loei Province.

Figure 8.12 The scaled model building task.

Engaging the hands in integrated STEM inquiry	Connecting to real world
Building the scaled model was more involved than what most students had thought. The task took quite a bit of time, and students began to appreciate the complexity of building pipeline systems in a mountainous region. The students also figured out different ways to build a scaled model so that they could get a more complete picture of the problem situation (See Figure 8.13).	Space for inquiry

114 *Growing humanistic values through integrated STEM inquiry*

Figure 8.13 Students building scaled model of the mountain.

After the model building, the attention was redirected to the problem and the ideating session began! Ms Kullawat wanted the students to ideate and design their solution based on what they had explored in the last two days. To facilitate the ideating and prototyping process, a bigger-sized scaled model of the mountains was provided (See Figure 8.14). Students were invited to examine the bigger-sized model and discuss how they would work out the design of their pipeline system.

Space for design and inquiry

Growing humanistic values through integrated STEM inquiry 115

Figure 8.14 Students discuss the design using the scaled model.

Although there was insufficient time for the students to build their actual prototype, the students realised the complexity of the problem when designing their pipeline system. Notably, the students realised that modelling real-world situations requires them to make many assumptions to simplify the problem. Simplifying the problem serves as a way for them to think about possible solutions during prototyping. An important realisation was the fact that while they could think about the pipeline system in terms of graph theory, there was an added challenge of the difference between a 'flat' graph and a 3D model. In other words, they would have to consider beyond the 'nice solutions' offered by mathematics and science and include practical considerations, which might not be optimal but feasible. This is an important learning point from these experiences!	Space for design, making, and inquiry Critique of ideas

8.2 Relating vignette to conditions for successful integrated STEM inquiry

This vignette draws on a real-world problem of remote villagers living in mountainous regions. Video clips with snippets of real villagers describing their way of life and the challenges they faced are used to develop students' empathy and to connect to students' curiosity. The use of videos featuring interviews with villagers is an important resource for bringing students to 'visit' remote areas. It is also a way for voices of different communities to be brought into classrooms. The use of videos is one way to introduce the sociocultural context in which the problem is situated, and this is an example of how the social aspect of the pedagogical practice of introducing the context and problem can enacted.

As it is unlikely that all students have direct experiences living in remote villages, creating time and space for students to raise their awareness of another living environment is important. To develop empathy, students must first understand, discuss, and reflect on factors affecting the lives of people living in remote villages and attempt to connect their new knowledge to their personal experiences.

The learning experiences were not merely focused on the material aspects of creating the system of pipes. Rather, time was allocated for direct teaching and learning of in-depth domain-specific knowledge related to graph theory, Bernoulli principles, scale, and map reading. The specific knowledge required is not developed in isolation but rather in the context of building the waterpipes. Claims and critique were also made in the context of building the waterpipes. For example, students had to consider the terrain of the mountains and the pressure of the water at different parts of the mountain, so it is not just about knowledge map reading and the concept of pressure. The importance of gaining the subject-matter knowledge to be applied to design the waterpipes cannot be underestimated. If students do not understand Prim's algorithm, mathematical modelling, or graph theory or Bernoulli's principles and fluid pressure, they would not be able to appreciate the contours, terrains, water pressure, and more importantly the challenges experienced by the villagers. Having such knowledge helps students focus on areas that are important in their design. This is an example of enactment of conceptual/procedural aspect of introducing the context and problem.

During the lessons, students were also provided with opportunities to engage actively with the material aspects of design, creating, and testing as they attempted to build the model. Through the experience of translating theoretical understanding to physical model, students appreciate the complexity of actual implementation in the mountains. As for evaluation and assessment of students' learning, vignette 3 highlighted the use of various tools such as graphic organisers, whole-class discussions, Thinking Board, modelling assignments, and hands-on investigations and challenges to allow students' ideas and thoughts to be made visible in public space. Teachers and students can subsequently engage with peer critique and feedback for idea improvement.

Table 8.1 maps out the variation in practice of the conditions for vignette 3. Students were presented with videos of villagers living in mountainous regions to enable them to develop an understanding of the challenges and constraints in accessing basic needs such as water. This way of engaging students' interests is typical of a user-centric STEM inquiry as the problem that is subsequently presented is one that is complex

Growing humanistic values through integrated STEM inquiry 117

Table 8.1 Understanding the variation in practice for vignette 3

Conditions	Problem-centric STEM inquiry	Solution-centric STEM inquiry	User-centric STEM inquiry
Connections to real-world problems	• Make connections to contemporary issues within specific local contexts. • Problems are presented as complex, persistent, and extended problems.	• Consider interactions between students and objects around them. • Present existing artefacts, technologies, or solutions to problems that are currently used in society.	• Interact with members of the community to understand their experiences and diagnose the problems they face.
Opportunities for presentation and critique of ideas and claims	• Apply domain knowledge related to science and mathematics to state and critique claims. • Critique also involved understanding the context in which the problem is located.	• Apply domain knowledge related to engineering and technology to critique design and workability. • Knowledge is also used to understand affordances, strengths, and limitations of current designs/solutions/ways of doing things.	• Apply knowledge of user needs and current social context to make claims and critique. • Use existing policies and rules to compare with the desired success criteria and user expectations.
Space for design, making, and investigation	• Students choose the materials they find most appropriate to design solutions. • Test solutions by carrying out experiments and using the data to re-examine the problem and solutions.	• Students work with existing prototypes/models to understand how they work and where changes/adaptations can be made. • Test out improved design to meet success criteria.	• Students consider ideas such as implementation details, accessibility, evaluation, and feedback mechanisms when users use product/solution.
Modelling communities of STEM problem solvers	• Work with experts (such as scientists or mathematicians) to learn techniques required to create solutions to the problems.	• Work with industrial partners or suppliers to understand existing product features, intentions, and limitations.	• Work with users in the community to understand different needs to make adaptations on the ground. Emphasis is on impact authenticity.

yet has practical importance. In this vignette, domain-specific knowledge that was applied spans across all the disciplines and hence showed characteristics of all three variants. In vignette 3, the students were not able to test their design on actual users to obtain feedback. However, they were able to test a model of the mountain, which also serves as good feedback. This is one way of enacting the pedagogical practice of reflecting on STEM learning experiences.

9 Working with evidence in integrated STEM inquiry

In all three variants, problem-centric, solution-centric, and user-centric STEM learning, collecting data and working with evidence is fundamental. Exposing students to different types of evidence (such as observations or measurements) will increase students' literacy of the relationship between the evidence and claims made. In vignette 4, we catch a glimpse of how Ms Tai transformed the familiar phenomenon of pigeon droppings into an integrated STEM inquiry lesson to involve students in designing ways to keep pigeons away from the school compound. Ms Tai and her colleagues planned activities for students to collect and use data to convince them that the large pigeon population in the school posed health threats. They also incorporated data collection as a means of evaluating the effectiveness of the prototypes developed.

The learning outcomes that Ms Tai and her colleagues strived to achieve from their STEM inquiry activity were for students to be able to: (1) explain the problems caused by the large population of pigeons living at their school; (2) explain the factors that encourage the pigeons to live in the school buildings; (3) research effective methods to repel the pigeons, while discussing and identifying the best way to design pigeon repellents; (4) design and build a model that indicates the position of each pigeon repellent based on The Geometer's Sketchpad (GSP); (5) assess the effectiveness of their design by placing the pigeon repellents at the designated buildings; and (6) present and discuss their results based on their proposals. This is an example of the epistemic aspect and pedagogical practice of clarifying the solution requirements and requesting for justification (s) in integrated STEM inquiry. Figure 9.1 shows how the learning intentions of Ms Tai and her colleagues' activity can be mapped onto the STEM quartet instructional framework.

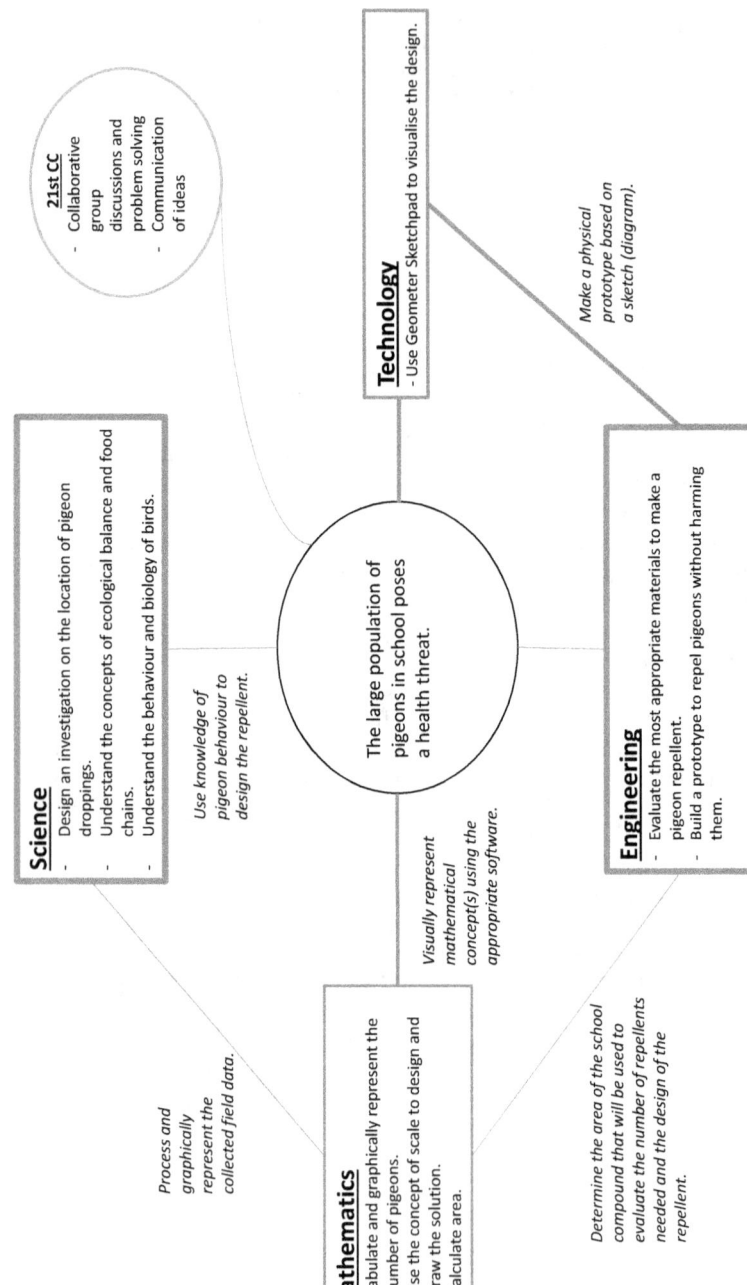

Figure 9.1 An integrated STEM lesson on designing a pigeon repellent using the STEM Quartet instructional framework.

9.1 Vignette 4—Harnessing data

Interactions in the classroom *Vignette 4: Ms Tai and students against pigeon droppings (adopting the problem-centric variant)*	Conditions observed
The problem: Pigeon Droppings Pose Health Threat Ms Tai and her team of teachers led a class of 40 students in an integrated STEM activity on solutions to repel pigeons from buildings in the school compound. The students were separated into eight groups consisting of five students per group. Students from each group were allocated roles and responsibilities, such as leader, note taker, poster lead, and presenter.	Connection to a real-world problem
Establishing the context The teaching team provided the context of the activity by bringing students on a walk around the school to look at the existing problems created by pigeons in each building in the school. Students observed the pigeons' behaviour and engaged in counting the pigeon populations in the designated school buildings. This one-week fact-gathering exercise supplied evidence for students to associate the large population of pigeons with the problems observed by students. For example, the information allowed students to identify the building with the highest number of pigeon visits and the time periods when the greatest number of pigeons were found. Through this exercise, students learned the skills of processing raw data and representing it graphically. Thereafter, students discussed and shared their opinions about the problems they observed by sharing their answers to open-ended questions on Mentimeter. The teachers then commented on the students' answers to scope and narrow down students' ideas to concretise the inconveniences caused by the pigeon population in school.	Space for investigation Opportunities for critique of ideas and claims
Next, students discussed and summarised the factors that encourage pigeon-dwelling in their school; this included access to abundant food, suitable dwelling structures, and the absence of natural predators. Each group of students organised their ideas in the form of a mind map using a flip chart and took turns presenting this information to the class (Figure 9.2). The presentation allowed students from different groups to share ideas and expand their knowledge. During the presentation, the teachers emphasised how reducing these factors may provide long-term solutions for controlling the pigeon population.	Opportunities for the critique of ideas and claims

Figure 9.2 Students presenting their mind map and ideas to their peers.

Learning about pigeon biology and ecology

After sufficient evidence was gathered to understand the scope and severity of the problems caused by the pigeons, the teachers guided the students to learn about the biology and ecology of pigeons, highlighting that knowledge of pigeon behaviour can help develop a more effective pigeon repellent.	Opportunities for presentation and critique of ideas and claims
Five learning stations were set up to enhance students' knowledge of pigeon biology and ecology. These stations included information on the pigeon's skeletal structure, vision, sound reception, odour perception, and behaviour, as well as the role pigeons play in an ecosystem. Examples of existing solutions used to repel pigeons were also showcased (Figure 9.3). Each of the five students in a group was assigned to learn the subject-matter knowledge from a different station. Upon completion, they returned to their respective groups to share what they had learned.	Space for investigation and critique of ideas

Working with evidence in integrated STEM inquiry 123

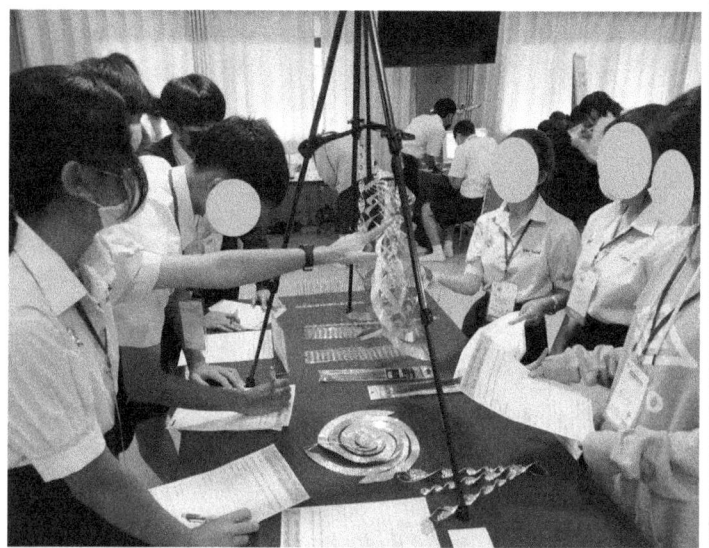

Figure 9.3 Students at a learning station on sound reception.

Ideating and prototyping

Before beginning the ideation process, the teachers reviewed the contents on pigeon biology and ecology and the information gathered on pigeon dwellings in the school buildings with the students. Each group then proceeded to brainstorm ideas and sketch their pigeon-repellent designs on flip charts. Their flip charts included information on the advantages and disadvantages of their proposal, how their ideas incorporated the information they had learned on pigeon biology and ecology, and details such as labels for the different design components, materials needed, and the cost of building their prototype (Figure 9.4).	Opportunities for presentation and critique of ideas and claims

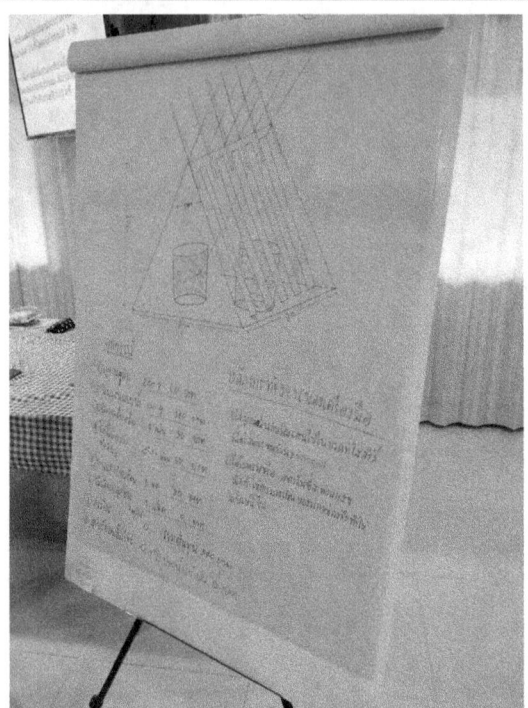

Figure 9.4 Flipchart showing the pigeon-repellent design proposed by a group of students.

The students then split into two large groups and presented their ideas using the hot seat technique (students are questioned by their peers) (Figure 9.5). During the presentation, each group received feedback from the remaining groups on how to improve their designs. Time was subsequently given to incorporate the comments from their peers and modify/improve their designs.

Space for design

Figure 9.5 Students getting feedback on their pigeon-repellent design from their peers.

Following initial design refinement, the students gathered the materials and tools required to build their prototypes. The materials were provided by the teachers and included pipes, water bottles, reflective discs, bird spikes, and tools (Figure 9.6). Some designs incorporated repelling scents and reflective discs, reflecting students' knowledge of pigeon sensory systems, while others were based on existing solutions, such as the inclusion of bird spikes.	Opportunities for presentation and critique of ideas and claims

Figure 9.6 Material and tools for constructing pigeon-repellent prototypes.

To guide the prototype-building process, students first used the software Geometer's Sketchpad to visualise their designs (Figure 9.7). Having had prior experience using this software, students incorporated mathematics with technology to produce visual representations (or models) to indicate the appropriate positions for their pigeon repellents. The model provided feedback on the structural feasibility of their design. In addition, students were able to apply knowledge of mathematical concepts, such as scale and geometry, as well as identify specific positions to place their pigeon repellents by calculating the area occupied by each repellent and the space between them. Guidance on how to use the materials and tools to model and construct the prototypes was provided by the teachers. Other calculations included the quantity of materials required and the cost of making the prototypes.

Space for design, making and investigation

Space for design, making, and investigation

Figure 9.7 Students using Geometer's Sketchpad to design a model of their prototype.

Students using Geometer's Sketchpad to design a model of their prototype

Upon completion of their designs on Geometer's Sketchpad, each group presented their prototype to the entire class (Figure 9.8). Guiding questions were provided to allow students to reflect on their solutions, evaluate their model-making ideas, and determine whether the criteria of suitability and appropriateness of the pigeon-repellent prototypes and their placement were met. The guiding questions included: Do you think the group's model can solve the pigeon problem, and why? How do you think the model will work if it is implemented in the school? Do you foresee any problems that may arise from this solution?	Space for design, making and investigation Modelling communities of STEM problem solvers

Figure 9.8 Students presenting their Geometer's Sketchpad model.

Towards the end of the activity, each group was given the opportunity to test their prototype by placing them in designated school buildings. Students were then expected to gather data for a period of one week to assess the effectiveness of their design in preventing or reducing the number of pigeon visits. The STEM activity ended with students reflecting on the activity as well as an explanation from their teachers on the significance of STEM education, to encourage them to think about how the knowledge and processes of STEM education can be applied to their daily lives.	Opportunities for presentation and critique of ideas and claims Space for investigation and connection to a real-world problem

9.2 Relating the vignette to conditions of successful integrated STEM inquiry

The lesson featured in vignette 4 incorporated several inquiry opportunities for students to engage with data collection and sense-making. In this vignette, the school compound acted as the familiar context within which the students' learning experiences were enacted. Given that the students go to school almost every day, they would likely have encountered and be familiar with the school's pigeon population. As such, it is also likely that they would have encountered the pigeon droppings! Intentionally connecting students' familiar experiences to the activity is important as students may have acclimatised to their surroundings, rendering the issue of the pigeon droppings 'invisible' to them. The observation task that requires the students to count and observe the distribution of the pigeons around the school hence offers a novel lens for students to experience a familiar phenomenon differently. Analysing the data collected helped convince the students that the growing pigeon population could potentially pose health problems in the future. Data collection and analysis can serve as ways to increase students' curiosity and interest in the issue at hand and encourage them to think about how problem identification is carried out in the real world. Being personally involved in the observation and data-gathering processes exposed students to how authentic field inquiry is carried out by scientists and researchers. The data also serves as evidence for students to understand the extent and scope of the problem created by the increasing pigeon population. Problematising familiar experiences requires deliberate efforts to 'see' what has become 'invisible' in the daily experiences of the students. This example highlights the social aspect and pedagogical practice of introducing the context and problem in integrated STEM inquiry.

Ms Tai and her team were also aware of the important role of domain-specific knowledge in facilitating student engagement with sensible and practical ideas to design their pigeon repellents. Understanding the biology and ecology of pigeons within the context of designing a repellent helps students relate the theoretical knowledge learned to a real-world application. Furthermore, using strategies such as the hot seat technique created opportunities for students to use the knowledge to communicate, persuade, and critique ideas. Through presentations, critique, and discussions, Ms Tai and her colleagues were able to assess the students' understanding of domain-specific knowledge and how they applied the knowledge to design their repellent. The strategic placement of the repellent around the school compound and data collection of the number of pigeons and the places they visited were powerful tools for evaluating the effectiveness of the students' prototypes. Ms Tai and her colleagues engaged with the pedagogical practice of clarifying the solution requirements and requesting for justification(s) through the social aspect of allowing students to make their ideas public for group critique and to persuade their peers of the soundness of their solutions.

Vignette 4 shows characteristics of problem-centric STEM inquiry with features of user-centric inquiry infused to create opportunities for feedback, refinement, and improvement. Students positioned their prototypes in various parts of the school to obtain data and inform them of areas that require refinement. The field test activity hence served to improve the students' solution through the processes of problem solving and design refinement (Table 9.1).

Table 9.1 Understanding the variation in practice for vignette 4

Conditions	Problem-centric STEM inquiry	Solution-centric STEM inquiry	User-centric STEM inquiry
Connections to real-world problems	• Make connections to contemporary issues within specific local contexts. • Problems are presented as complex, persistent, and extended.	• Consider interactions between the students and objects around them. • Present existing artefacts, technologies, or solutions to problems that are currently used in society.	• Interact with members of the community to understand their experiences and diagnose the problems they face.
Opportunities for presentation and critique of ideas and claims	• Apply domain knowledge related to science and mathematics to state and critique claims. • Critique also involves understanding the context in which the problem is located.	• Apply domain knowledge related to engineering and technology to critique design and workability. • Knowledge is also used to understand affordances, strengths, and limitations of current designs/solutions/ways of doing things.	• Apply knowledge of user needs and current social context(s) to make claims and critique an idea/a product/solution. • Compare existing policies and rules to the desired success criteria and user expectations.
Space for design, making, and investigation	• Students choose the materials they find most appropriate to design solutions. • Students test solutions by carrying out experiments and using the data to re-examine the problem and the solutions.	• Students work with existing prototypes/models to understand how they work and where changes/adaptations can be made. • Students test out their improved design against success criteria.	• Students consider ideas such as implementation details, accessibility, evaluation, and feedback mechanisms when users use product/solution.
Modelling communities of STEM problem solvers	• Students work with experts (such as scientists or mathematicians) to learn techniques required to create solutions to the problems.	• Students work with industrial partners or suppliers to understand existing product features, intentions, and limitations.	• Students work with users in the community to understand different needs to make adaptations on the ground. • Emphasis is on impact authenticity.

10 Using expert voices to increase task authenticity

The integrated STEM inquiry featured in vignette 5 was planned as a user-centric STEM activity on bettering the lives of rubber tappers through improving the design of latex storage containers. The overarching goal of the STEM challenge was closely tied to the major economic and health concern of how the daily work of rubber tappers in the community can be made less stressful or laborious. Unique to this activity was the presence of expert voices to give students authoritative feedback and hold them accountable for their ideas and designs.

Rubber tapping, or the extraction of latex in its raw form, is an extremely labour-intensive process even in this modern era. It requires human workers to expertly cut shallow, sloping semi-circular rings on the bark of rubber trees every morning (approximately 0200–0600 hours), and then return to these trees to collect the exuded sap or latex a few hours later. While each tree might only produce a small cupful of liquid latex, this volume quickly accumulates as the rubber tapper makes his or her rounds. Currently, the latex is carried around in a container strapped onto the tapper's back, who then brings it to a collection point in exchange for wages based on the amount collected. This is backbreaking work, and rubber tappers suffer from various health hazards from physical labour, the side effects of the chemicals from rubber-making process, and even dangers of being attacked physically by wild animals. As a cash crop, however, rubber production has made Thailand the number one producer of latex in the world and the province of Trang the major supplier for the country. It has been the main reason for the economic prosperity of the people in the region and crafted as a worthy STEM challenge for students in this school.

Mr Tub is the teacher featured in this vignette, and he worked with his colleagues to plan a lesson that aimed at increasing students' awareness of the plight of rubber tappers. They also invited experts to their classrooms to increase the authenticity of learning experiences. The map for their lesson plan is featured in Figure 10.1.

DOI: 10.4324/9781003422501-10

132 Using expert voices to increase task authenticity

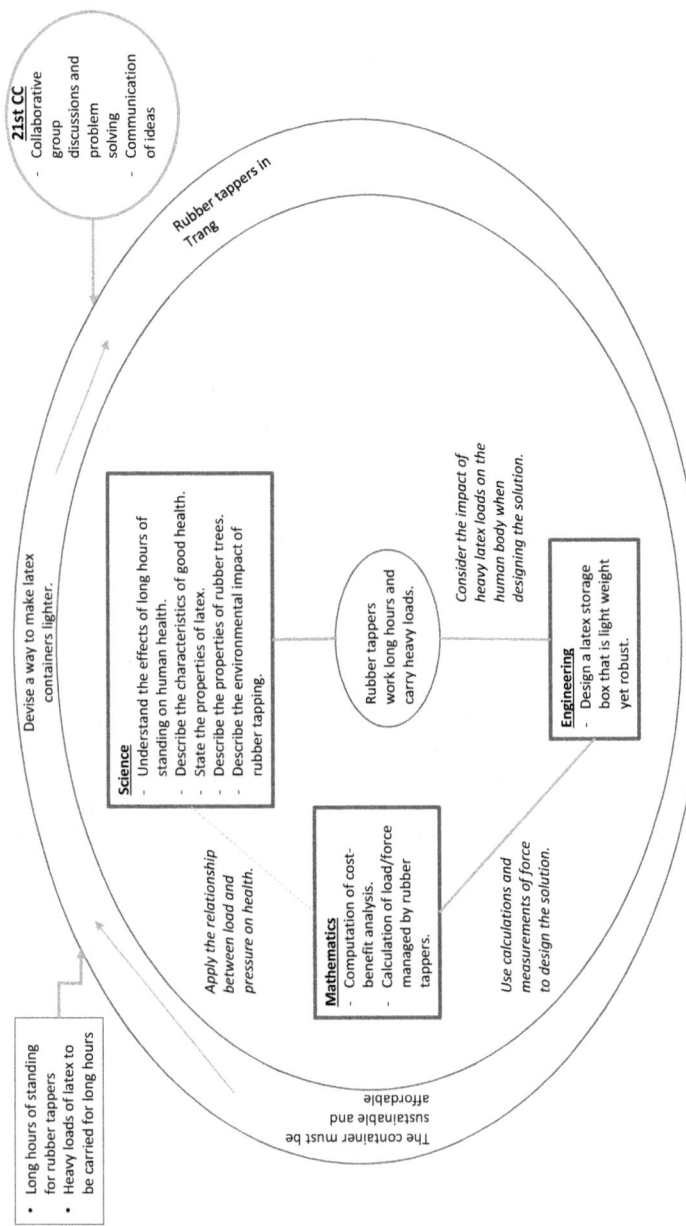

Figure 10.1 An integrated STEM lesson on improving the lives of rubber tappers using the STEM Quartet instructional framework.

10.1 Vignette 5—Improving lives

Interactions in the classroom	Conditions observed
Vignette 5: Improving the lives of rubber tappers (adopting problem-centric and solution-centric variants)	
The problem: Reducing the load carried by rubber tappers	
Mr Tub and his team of teachers planned an integrated STEM inquiry on innovating existing methods of latex storage to alleviate the difficulties experienced by rubber tappers when carrying a heavy load. Through this lesson package, the team hopes that students can identify the everyday problems faced by rubber tappers, develop a greater understanding of the health issues they encounter, and conduct research to learn about innovative latex storage solutions and evaluate the existing solutions for latex storage.	Relating to real-world context
Mr Tub started the activity by having students search for information on the physical features of rubber plantations, the chemical properties of latex, and the processing of rubber tapping. Students watched a few video clips on rubber tapping and environmental protection and read relevant research articles concerning the health issues confronted by the industry.	Space for investigation
Armed with the necessary background knowledge, the teachers and students discussed how latex harvesting has affected the health of rubber tappers and collectively agreed on the features essential for effective storage of latex. The challenge was to design a new model or innovation for latex storage. The solutions would then be evaluated on the following conditions: (1) does not affect the health of workers adversely, (2) is cost-effective, (3) is able to carry heavy loads and has automated movement, and (4) uses recycled or biodegradable materials.	
Identifying the problem and ideating solutions	
Upon understanding the context and challenges faced by rubber tappers, the students started group discussions to scope the problem they would like to address. Some groups began listing the health problems associated with rubber tapping, while others examined the advantages and disadvantages of current latex storage devices. Students listened to the ideas presented during group discussions and provided intra-group feedback to improve their draft model/innovation. To obtain further input on their solutions, students presented their draft designs to other groups for peer critique and evaluation. Invited experts from nearby universities also gave the students advice and feedback. This increased the authenticity of the challenge and placed further emphasis on the activity's connection to the real world.	Opportunities for critique of ideas and claims

After several rounds of feedback and refinement from different groups and experts, the students tested their models/innovations by filling them with water and presenting their final model/innovation to the class. The following day, the teachers brought liquid latex from the plantations for the students to test their models/innovations by simulating the real-world situation and collecting data as evidence of the feasibility and usefulness of their solutions. Some of the outcomes (data) measured included the volume of latex collected and the speed of the collection process. This was a very exciting part of the activity and although it took nearly half a day, each group was thoroughly engaged in the solution-testing process and took the challenge very seriously. The challenge concluded with each group presenting their findings to their classmates followed by a class discussion on the pros and cons of their models/innovations.	Modelling communities of STEM problem solvers Connection to a real-world problem Presentation and critique

10.2 Relating the vignette to the characteristics for successful integrated STEM inquiry

Students were presented with the authentic real-world context of the physical and health-related challenges faced by rubber tappers. This context may be unfamiliar to students as they may not have interacted with rubber tappers before. To cultivate their awareness and interest, the teachers created learning opportunities through tasks such as watching video clips for students to learn about the science of rubber tapping and better understand the lives of rubber tappers and the challenges they face. The attention paid to developing in-depth, domain-specific, and context-specific knowledge served as a primer for students to delve further into their research and look for innovative ways to design their latex storage solutions. Once they acquired sufficient domain-specific knowledge, the students were better able to engage with one another to construct higher-quality claims and critique each other more convincingly. Through many rounds of presentations and feedback from both their peers and experts, students developed competencies to present, persuade, and critique.

The university experts added authenticity to the activity in two dimensions. First, students were held accountable for their ideas as they explained and justified their solutions to experts who provided authentic feedback. Second, the questions and feedback from experts served as models for students to develop noticing abilities of experts. By listening to the questions asked and understanding the different aspects of feedback provided, students (and teachers) get an idea of what experts value when evaluating latex storage systems. Comments from experts are an ideal way of engaging students in the epistemic aspect of justifying their decisions and decision-making processes in integrated STEM inquiry.

Vignette 5 offers an interesting hybrid way of enacting STEM inquiry. While the teachers started with the intention to improve the lives of rubber tappers by presenting students with information on the challenges they face (problem-centric), their dual focus became evident in the design activity, which sought to improve the current ways of storing latex (solution-centric). As such, when mapping the inquiry practices to the conditions of integrated STEM inquiry, we observed connections to real-world problems, disciplinary-specific knowledge, practices, and competencies—characteristics of both solution-centric and user-centric inquiry (Table 10.1).

Table 10.1 Understanding the variation in practice for vignette 5

Conditions	Problem-centric STEM inquiry	Solution-centric STEM inquiry	User-centric STEM inquiry
Connections to real-world problems	• Make connections to contemporary issues within specific local contexts. • Problems are presented as complex, persistent, and extended.	• Consider interactions between the students and objects around them. • Present existing artefacts, technologies, or solutions to problems that are currently used in society.	• Interact with members of the community to understand their experiences and diagnose the problems they face.
Opportunities for presentation and critique of ideas and claims	• Apply domain knowledge related to science and mathematics to state and critique claims. • Critique also involves understanding the context in which the problem is located.	• Apply domain knowledge related to engineering and technology to critique design and workability. • Knowledge is also used to understand affordances, strengths, and limitations of current designs/solutions/ways of doing things.	• Apply knowledge of user needs and current social context(s) to make claims and critique an idea/a production/solution. • Compare existing policies and rules with the desired success criteria and user expectations.
Space for design, making, and investigation	• Students choose the materials they find most appropriate to design solutions. • Students test solutions by carrying out experiments and using the data to re-examine the problem and the solutions.	• Students work with existing prototypes/models to understand how they work and where changes/adaptations can be made. • Students test out their improved design against success criteria.	• Students consider ideas such as implementation details, accessibility, evaluation, and feedback mechanisms when users use product/solution.
Modelling communities of STEM problem solvers	• Students work with experts (such as scientists or mathematicians) to learn techniques required to create solutions to the problems.	• Students work with industrial partners or suppliers to understand existing product features, intentions, and limitations.	• Students work with users in the community to understand different needs to make adaptations on the ground. • Emphasis is on impact authenticity.

11 Conclusions

Chapters 6 to 10 presented different ways of enacting integrated STEM learning in the classroom. In some classrooms, the focus was on the material aspects of STEM learning, where students designed and constructed prototypes. In other classrooms, more attention was paid to developing students' domain-specific knowledge through problem solving. The diversity of practices in integrated STEM inquiry is not unexpected. As problems chosen by teachers are dependent on local contexts, and the nature of problems influence the STEM inquiry experiences, by inference, integrated STEM inquiry is context-sensitive and dependent. The teachers who participated in this study attended a three-day symposium and an online learning module focusing on planning integrated STEM learning experiences using the STEM Quartet instructional framework. They also worked with a team of international consultants to refine their lesson ideas to better align with ideas of integration across subject disciplines. More details of the programme can be found at http://www.meristem.site.

Despite differences in the problems presented, the five vignettes also showed common practices. For instance, students in all of the classrooms worked with data, raised questions, stated claims, critiqued ideas, and actively worked on constructing prototypes. These similar practices define the experience of integrated STEM inquiry as distinct from monodisciplinary learning of science or mathematics.

For the rest of this chapter, we present ten key observations about integrated STEM inquiry from the ideas discussed in Chapters 1 to 5 and from the vignettes described in Chapters 6 to 10.

Observation 1: Integrated STEM inquiry learning experiences involve students engaged with complex problems or issues that are found in the real world. Immersing themselves with real-world problems or issues helps to build students' domain-specific knowledge and develop their 21st-century competencies.

Observation 2: Time and opportunities must be set aside for students to learn domain-specific knowledge and skills to engage in specific integrated STEM inquiry. Explicit instruction of domain-specific concepts may be needed, especially when the problem/issue presented to students is novel (Sewell, 2021). This is to avoid randomness in identifying or understanding the problem/issue. Randomness or trial and error in integrated STEM inquiry could result in a waste of time and resources and augment the frustration experienced by learners. As such mapping the intended learning outcomes to the problem/task/issue is necessary. The three variants of the STEM Quartet instructional framework form a useful tool for mapping the learning outcomes.

Observation 3: Integrated STEM inquiry learning experiences must include opportunities for students to ask questions, design investigations, make observations, and collect data. This ensures that students can make evidence-informed arguments, explanations, and decisions.

Observation 4: Integrated STEM inquiry is a good platform for students to develop their 21st-century competencies of collaboration, communication, and inventive thinking. Through interactions with their peers and teachers, students learn to present their ideas, persuade others to adopt them, and work productively in teams. These social interactions mirror how communities of problem solvers work in the real world.

Observation 5: There are different ways to evaluate students' learning in integrated STEM inquiry. The evaluation methods selected should be aligned with the intended learning outcomes of the activity. Different aspects of the STEM inquiry experience can be evaluated. Evaluation can focus on conceptual understanding within specific domains, workability of the prototype created, number of feasible ideas generated, or how group members work together.

Observation 6: During integrated STEM inquiry, students spend more time working on the assigned task and engage with more activities that are directed by them selves. Teachers need to be mindful to set aside time to support students in exploring, collecting data, making refinements, discussing, and reflecting. Consequently, more curriculum time is needed to enact STEM inquiry.

Observation 7: Prototypes or models created as part of the solutions of integrated STEM inquiry may not always be of high quality. The variability in quality could be due to student competencies in making, the type of materials provided, or the amount of time available. It is unrealistic to expect only high-quality workable prototypes from the limited class time available, since problem solvers can spend many months or years developing their products in the real world.

Observation 8: It is unlikely that the solutions generated from integrated STEM inquiry experiences in school will be novel or ground-breaking. It is likely that the solutions proposed are adaptations of common inventions or objects in the students' immediate environment. Students could get ideas from searching the World Wide Web to understand current solutions before adopting them. A solution that is not novel does not make the students' learning experience any less valuable.

Observation 9: The choice of which variant (problem-, solution-, or user-centric) of the STEM Quartet instructional framework to use is dependent on your learning outcomes and your learner profile. For younger learners who are experiencing integrated STEM inquiry for the first time, consider using solution-centric inquiry to scope their investigation. For more advanced learners, a problem-centric approach would be more meaningful. The user-centric approach would be appropriate if you intend to emphasise ideas of inclusion and diversity in your lessons.

Observation 10: While most solutions proposed by students are equally plausible, not all suggestions are equally defensible. As such, getting students to justify their ideas, explain the rationale of their decisions, and consider the trade-offs they made is important. The space for peer or expert critique is a rich learning platform for all students.

We hop these 10 observations would be useful references for teachers who are beginning their integrated STEM journey. For STEM educators and researchers, the key observations presented in this book could form the basis of your courses for teachers or agendas for further research.

References

Sewell, J. (2021). *Why inquiry-based approaches harm students' learning.* Analysis paper 24. The Centre for Independent Studies.

Index

Pages in *italics* refer to figures and pages in **bold** refer to tables.

21st Century 1–2, 26, 55, 102; competencies 2, 11, 22, 26, 48, 51–52, 54–55, 57–63, 65, 70, 81, 100, 134–135, 137–138; competency 58, 60, 62–63, 79, 81
4Cs 77; Collaboration 2, **39**, 52, 65, 69–71, 75, 79, 85, 102, 138; Communication 2, 10, **39**–40, 52, 54, 58, 60, 62, 68–69, 75, 79–81, *93–94*, 102–*103*, 109, *120*, *132*, 138; Creativity 2, 25, 32, 36–37, 40–41, 46, 52, 54, 58, 60–62, 64, 68; Critical thinking 20, 40, 52, 54, 58, 60, 62, 64, 102
5E model 14–15, 17
5Es: Elaborate 14, 46, 70; Engage 8, 11, 13–16, 29, 31, 33–34, 37, 43–**44**, 47–52, 57, 59, 65, 73–74, 90, 99, 102, 109, 116, 128, 134, 137–138; Evaluate 14, 33–34, 52, 54, 64–65, 69–70, 74, 79, 84, 90, 92, *120*, 127, 133, 138; Explain 5, 10, 13–14, 22, 50, *80*, 97, 119, 139; Explore 6, 12, 14, 21, 83, 86, 92

Ability to reflect 2
ABS 99
Acquisition 48, 81
Actions 9–10, 42, 46, 89
Activity 21, *23*, 25, 37–38, 48–51, 58–59, 61–62, 65, 70, 76, 81, 89–90, 99–100, 102, 109, 119, 121, 128–129, 131, 133–135, 138; activities 7, 14, 27–28, 32, 38–**39**, 41, 47–48, 57, 59, 61, 63, 65–67, 71, 74, 79, 95, 102, 119, 138; creative 11, 20, 32, 37, 41, 60, 66, 69, 75, 95
Adaptations **53**–54, **77**, **91**, **101**, **117**, **130**, **136**, 138
Advice 51, 95, 133
Affective 57, 60
Affordances 7, 15, 22–*23*, 27, 31, **53**, **77**, **91**–92, **101**, **117**, **130**, **136**; advantages 22, 31, 92, 123, 133; disadvantages 22, 31, 92, 123, 133; pedagogical 7, 11–12, 15, 21–22, 26, 29–30, 43–45, 72, 89, 99, 116, 118–119, 129
Analyse 35, 38–**39**, 61; analysis 26, 30, 38–40, 62, 129, *132*, 139
Apparatus 51–52, 96, 98
Application 5, 7, *23*, 28, 42, 46, 49, 52, 68, 76–**77**, *103*, 106, 129
Applications 11, 75, 92
Approach 6–7, 11, 14, 17, 26, 37, 48, 92, 99, 139; approaches 6–7, 24, 26–27, 29, 36, 139; inquiry-based 7–8, 11–12, 14–15, 25, 55, 106, 139; investigative 10–11, 27, 51; pedagogical 7, 11–12, 15, 21–22, 26, 29–30, 43–45, 72, 89, 99, 116, 118–119, 129
Arguments 5, 7–8, 10, 138; logical 5, 7, 10, 12–14, 28, 42, 52
Artefacts **53**, 75, **77**, 81, **91**, 95, 99–**101**, **117**, **130**, **136**
Assess 21, 57–61, 63–66, 68, 70, 90, 119, 128–129
Assessment 57–72, 116; competence-based 58–60, 62, 64–65; integrated

Index 141

STEM 2, 4, 6, 9, 11–15, 17–18, 21–79, 88–89, 99–100, 102–*132*, 134–135, 137–139; models 2–4, 6, 9, 12, 14–18, 25–26, 29–31, 36, **39**, 51–**53**, 55, 57–58, 62, 70, **77**, **91**, **101**, **117**, 126, **130**, 134, **136**, 138; observational 19–20, 64; strategies 38, 46, 48–50, 67, 70–71, 79, 81, 129; student self-reported 63–65, 67; teacher-subjective 63–67, 69
Attitude 60, 64–65
Authentic 8, 12, 15, 18–19, 31, 47–48, 54, 56, 63, 76, 129, 134; authenticity 47–49, **53**, **77**, **91**, **101**, **117**, **130**–**136**
Authoritative 52, 131
Awareness 2, 104, 116, 131, 134; inter 2, 43, 72; intra 2, 133

Background knowledge 12, 48, 133
Basin 98–99; design *3*, 5, 13–15, 18, 20, 22–*23*, 25–26, 28–33, 35, 37, **39**, 43, 49–50, 52–**53**, 55, 62, 67, 69, 71–72, 75, **77**, *80*–81, 83–85, 88, 90–**101**, *103*, 108, 110, 114–*120*, 123–**136**, 138; large 51, 73, 98–99, 119–121, 124
Bereiter 20–22, 24, 47, 54
Bernoulli's principle 104, 111
Biology 5, 20, *120*, 122–123, 129
Biomimicry 92–95, 99–100
Blades 92, 96; brittle 96; broken 76, 96; turbine 92
Blocks 98–99; prepared 98, 106, 112
Box cutters 96
Brainstorming session 98
Budget planning 100
Build 2, *23*, 45–46, 48, 50, 52, 61, 82, 84, 86, 89–90, 112–113, 115–116, 119–*120*, 125, 137; hands-on building 81; pipeline system 110, 112, 114–115

Cause and effect 20, 38
Challenge 7, 24, 31, 95, 97–99, 108, 110, 115, 131, 133–134; challenges 1–2, 10, 21, 26, 47, 66, 71, 75, 88, 104–105, 115–116, 133–135; foldable raft challenge 98
Changes 1–2, 20, 37–38, 51, **53**, 76–**77**, **91**, **101**, **117**, **130**, **136**

Characteristics 2, 4–5, 9, 15, 29–30, 42, 46, 61–62, 72, 76, 90, 92, 100, 118, 129, *132*, 134–135
Check 43, 65–66, 68, 74, 83, 87
Cheerful music soundtrack 95
Chemistry 5–6
Childhood games 99
Claims 5, 7, 43–**44**, 49–50, **53**, 72, **77**, 82, 89, **91**, **101**, 116–**117**, 119, 121–123, 125, 128, **130**, 133–134, **136**–137
Clarifying 7, **44**, 73–74, 89, 119, 129
Class 22, 25, 59, 69, 82, 89–90, 99, 105–106, 110–111, 116, 121, 127, 134, 138; classmates 134; classroom 10, 25, 27, 36, 38, 40–41, **44**, 55, 59, 64, 70–73, 75–76, 84, 86, 121, 133, 137; classrooms 8, 24–25, 27, 32, 43, 54, 72–78, 82, 95, 99, 104, 115, 131, 137; practices 2, 5–6, 8–9, 11–15, *23*–24, 27–31, 33, 35–45, 48, 54, 57–58, 60–61, 64–66, 68–73, 89, 92, 102, 135, 137; time 8, 15, *23*, 32–33, 38, 50–51, 69, 73, 79–81, 87, 95, 98–99, 108, 110, 113, 115–116, 121, 124, 137–139
Claws 96–97, 99; covered 20, 96; crab *93*, 97–99
Clips 104, 111, 115, 133–134; video 100, 105, 111, 115, 133–134
Coding 79–81, 90
Cognitive 4, 27, 32, 40, 42, 47, 61–62
Collaboration 2, **39**, 52, 65, 69–71, 75, 79, 85, 102, 138; critiqued 137
Collecting 8, 20, 43, 74, 119, 134, 138
Comments 69, 124, 135
Communicating 29–30, 95
Community 18, 42, **44**–**45**, **53**, 74, **77**, 90–**91**, 100–**101**, **117**, **130**–131, **136**
Competency 58, 60, 62–63, 79, 81; competencies 2, 11, 22, 26, 48, 51–52, 54–55, 57–63, 65, 70, 81, 100, 134–135, 137–138; domain-specific competencies 54, 58
Competitive 90
Components 31, 34, 59–60, 63, 96, 123
Computation 11, *23*, *132*; computational thinking 9, 29, 41, 54, 60, 79; computational thinking skills 60; computational thinking-focused 29
Concept 8–9, *23*, 32, 36, 49, 69, *80*, 83–84, 89, 92–*93*, 99–100, *103*, 110,

142 Index

116, *120*; conceptions 24, 26, 42, 46, 48; concepts 5, 10, 13, 16, 27–28, 38–**39**, 41, 45–46, 48–49, 58, 71, 79–*80*, 82, *103*, 106, 110, *120*, 126, 138; conceptual *3*–4, 6–8, 11–12, 14, 17–18, 25, 28, 30, 32, 43–45, 47, 56–57, 60, 73, 75–76, 90, 100, 116, 138; conceptual understanding 4, 28, 57, 138; domain-specific concepts 138; personal conceptions 46; preconceptions 48
Conditions 19–21, *23*, 37–38, 43, 46, **53**–54, 72, 76–**77**, 82, 88, **91**, 95, 99, **101**, 104, 111, 115–**117**, 121, 128, **130**, 133, 135–**136**
Conjecturing 9–10, 13, 28, **44**
Connect 14, 20–21, 33, 37, 43–**44**, 49, 54, 58, 65, 68, 73, 75, *80*, 115–116; connections 4, 15–17, 22–*23*, 31, 34, 46, 49, **53**–54, 58, 73–74, **77**, 79, 83, **91**–92, **101**, 109, **117**, **130**, 135–**136**
Cons 134
Constraints 5, *23*, 27, **44**, 46, *80*, 89, 116
Constructed 22, 69–70, 137
Contemporary issues **53**, 73, **101**, **117**, **130**, **136**
Content knowledge 36, 38, 49, 81
Contents 123
Context *3*–4, 9, 11, 13, 19, 31–33, 35–36, 43–**44**, 46–49, 52–**53**, 58–59, 61–63, 65, 73, 75–**77**, 82, 89–**91**, 99, **101**–102, 104, 116–**117**, 121, 128–**130**, 133–134, **136**–137; contextualising learning 102; curricular 4; local context 75; local contexts **53**, 73, **77**, **91**, **101**, **117**, **130**, **136**–137; real-world *3*, 14–16, 18–19, 27–28, 31, 33, 42–**44**, 46–49, **53**, 57, 72–73, **77**, 82, 88–**91**, 99, **101**–102, 106, 115, **117**, 121, 128–**130**, 133–137; sensitive 31, 40, 64, 137; specific 2, 4, 6, 10–11, 13–15, 22, 31, 33–35, 38, 41–43, 45–47, 49–50, 52–54, 58–59, 61, 64, 68, 73–79, 81, 84, 89–**91**, 99, **101**–102, 111, 116–118, 126, 129–**130**, 134–138
Convince 119, 129
Crab *93*, 97–99; freshwater 98
Creative 11, 20, 32, 37, 41, 60, 66, 69, 75, 95; medium 11
Credits 95, 100

Criteria 14, 30, 32, 41, **44**, 50–51, **53**, 59, 64, 66, **77**, 81–82, 89–**91**, 100–**101**, **117**, 127, **130**, **136**; success 12, 14, 32, **44**, 50, 52–**53**, **77**, 81–82, 89–**91**, 100–**101**, **117**, **130**, **136**
Critical 17, 20–21, 34, 40, 52, 54, 58, 60, 62, 64, 69–70, 102, 110; mathematical modelling 20, 76, 109, 116; thinking 8–10, 14–15, 18, 20, 22, 25–27, 29–30, 40–41, 52, 54–55, 58, 60, 62, 64, 66, 74, 79, 95, 102, 105–106, 109, 116, 138
Critique 22, 43–**44**, 49–50, **53**, 68, 72–74, **77**, 82, 88–**91**, 95, 97–99, **101**, 105, 108–110, 115–**117**, 121–123, 125, 128–**130**, 133–134, **136**, 139
Curiosity 7, 19, 48, 115, 129
Curriculum 8, 24, 26–27, 41, 58, 71, 78, 138

Data 8–9, 12–13, 20, 29–30, 37–40, 43, 51–**53**, 61–64, 73–**77**, *80*–81, 90–**91**, 99, **101**, **117**, 119–121, 128–**130**, 134, **136**–138; collection 31, 51, 119, 128–129, 131, 134; gathering 9, 73, 121, 129
Debates 8, 33, 43, 49–50
Decision **44**, 59, 72, 81, 100, 135; decisions 2, 22, 33, 36, 42, **44**, *80*, 90, 135, 138–139
decision making 72, 100
Deck surface 98
Deductive reasoning 10–11
Defining 4, 9, 29–32, 42, 51, 55, 68
Design *3*, 5, 13–15, 18, 20, 22–*23*, 25–26, 28–33, 35, 37, **39**, 43, 49–50, 52–**53**, 55, 62, 67, 69, 71–72, 75, **77**, *80*–81, 83–85, 88, 90–**101**, *103*, 108, 110, 114–*120*, 123–**136**, 138; dominant domain knowledge 75; educational approach 6; educational curricula 28; educational environments 4; educational policies 64; in-depth 89, 112, 116, 134; knowledge 2, 4–24, 28, 33–34, 36, 38, 40, 42, **44**–50, 52–54, 57–65, 68, 73–**77**, 79, 81, 89–**91**, 100–102, 108, 112, 116–118, *120*–122, 125–126, 128–**130**, 133–137; specific 2, 4, 6, 10–11, 13–15, 22, 31, 33–35, 38, 41–43, 45–47, 49–50, 52–54, 58–59,

Index 143

61, 64, 68, 73–79, 81, 84, 89–**91**, 99, **101**–102, 111, 116–118, 126, **129**–**130**, 134–138
Development 26–27, 36–37, **44**, 46–47, 50, 54–56, 59, 63, 67–69, 71, 73, 81, 88, 100, 102, 108
Devices 5, 48, 61–62, 99, 133
Diagnose **53**, **77**, **91**, **101**, **117**, **130**, **136**
Difference 14, 99–100, 115
Direction 96
Disciplinary 4–6, 8–17, 22–24, 27–30, 33, 36–37, 42–46, 49–50, 55, 57–59, 61, 63, 65, 68–69, 71, 74, 79, 90, 102, 135; cross-disciplinary 36; features 4, 6, 16–17, 19, 28, 31–32, 42–43, **53**, 61, **77**, **91**, **101**, **117**, **129**–**130**, 133, **136**; goals 4–6, 14–15, 29, 31, 38, 47, 51, 56, 59, 65, *80*, 88; interdisciplinary 16, 26, 28, 57–58, 62–63, 71, 79; integration 4–7, 13, 15–16, 28, 43–**44**, 60, 63, 71, 137; practices 2, 5–6, 8–9, 11–15, *23*–24, 27–31, 33, 35–45, 48, 54, 57–58, 60–61, 64–66, 68–73, 89, 92, 102, 135, 137; knowledge 2, 4–24, 28, 33–34, 36, 38, 40, 42, **44**–50, 52–54, 57–65, 68, 73–**77**, 79, 81, 89–**91**, 100–102, 108, 112, 116–118, *120*–122, 125–126, **128**–**130**, 133–137; monodisciplinary 43–45, 57, 63, 137; multidisciplinary 16, 20, 79–**91**; practices 2, 5–6, 8–9, 11–15, *23*–24, 27–31, 33, 35–45, 48, 54, 57–58, 60–61, 64–66, 68–73, 89, 92, 102, 135, 137; skills 2, 9, 13, 15–17, 20, 22, 26–29, 31, 33–34, **45**, 47, 49, 52, 54–55, 57–58, 60–62, 64–65, 68–69, 73–75, 79–82, 89–90, *93*–*94*, 96, 102, 121, 137; transdisciplinary 15–16, 18–19, 57–58, 79; understandings 14, 48, 57, 73
Discipline 2, 4–8, 10, 16, 33, 35, 37–38, 42, **44**; boundaries 4, 42–43; disciplines 2–9, 12–17, 19–20, 22–24, 28–29, 33–35, 37–38, 42–43, 48–50, 58, 74, 79, 102, 118, 137; multiple 2, 4, 7, 20, 31, 46, 59, 64, 74, 79
Discuss 4, 13, *23*, 32, 57, 59, 81, 90, 108, 114–116, 119; discussing 84, 119, 138; discussion 8, 28, 33, 68, 73, 99–100, 105–106, 111, 134; group discussions 49, 65, 81, 89, *93*–*94*, 133
Diversity *80*, 137, 139
Drawing 12, **39**, 63, *80*
Drilling 81, 85, 89
Drinking straws 96

Earth 5
Ecology 122–123, 129
Education 2–4, 6–8, 14, 18, 24–29, 41, 52, 55–56, 59, 64, 70–71, 78, 128; educators 7–9, 14, 27, 139; experimenting 8; guided 11, 14, 20, 27, 51, 73, 75, 81, 90, 111, 122
Effectiveness 14, 46, 74, *80*, 119, 128–129
Empirical 5, 8–9, **44**, 51, *80*, 89; data 8–9, 12–13, 20, 29–30, 37–40, 43, 51–**53**, 61–64, 73–**77**, *80*–81, 90–**91**, 99, **101**, **117**, 119–121, 128–**130**, 134, **136**–138; evidence 5, 8–9, 12–14, 21, 29–32, **44**–**45**, 49–51, 73, 76, 81, 90, 100, 119–**130**, 134, 138
Empowering 27, 73; mathematics 2–5, 7, 9–12, 14, 17–18, 20, 22–*23*, 25–30, 35, 37, **39**–40, 42, **44**, 49–51, **53**, 55, 71, 73–74, 76–**77**, 79–81, 90–*94*, **101**, *103*, 106, 108–109, 115, **117**, *120*, 126, **130**, *132*, **136**–137; science 2–10, 12–15, 17–18, 20, 22–31, 35, 37–43, 48–51, **53**, 55–56, 60, 71, 73–74, 76–81, 89–*94*, **101**, *103*, 115, **117**, **130**, *132*, 134, **136**–137; STEM 1–*80*, 82, 88–92, 99–139
Enact 31, 76, 138; enacted 38, 41, 51, 54, 71–72, 116, 128; enacting 40, 43, 48–50, 118, 135, 137; enactive thinking 14–15; enactment 40, 42–45, 72–78, 81, 90, 102, 116
Engage 8, 11, 13–16, 29, 31, 33–34, 37, 43–**44**, 47–52, 57, 59, 65, 73–74, 90, 99, 102, 109, 116, 128, 134, 137–138
Engagement 8, 22, 32, 43, 45–46, 48–49, 52, 55–56, 71, 74, 89, 129; cognitive 4, 27, 32, 40, 42, 47, 61–62; student 7, 13, 21, 24–25, 31, 38, **45**, 52, 54, 58–67, 70–71, 81, 97, 100, 129, 138
Engaging 7, 17, 29–31, 45, 65, 74, 104, 108, 113, 116, 135

144 Index

Engineering 2–5, 9, 12, 14, 17–18,
  22–24, 26–31, 35, **39**–41, 49–50, **53**,
  60–62, 75, **77**, 79–81, 89–*94*, **101**,
  103, **117**, *120*, **130**, *132*, **136**; design
  3, 5, 13–15, 18, 20, 22–*23*, 25–26,
  28–33, 35, 37, **39**, 43, 49–50, 52–**53**,
  55, 62, 67, 69, 71–72, 75, **77**, *80*–81,
  83–85, 88, 90–**101**, *103*, 108, 110,
  114–*120*, 123–**136**, 138; focused 10,
  29–30, 35, 58, 62, 100, 102, 111,
  116; ideas 2, 4, 6–8, 10–12, 14–15,
  22, 29–30, 32, 36–37, 42–50, 52–55,
  58–59, 61–63, 65–66, 68–70, 72–74,
  76–**77**, *80*, 82, 88–89, **91**, *93*–95,
  97–**101**, *103*, 105–106, 108–110,
  115–**117**, *120*–125, 127–134,
  136–139; literacy 18, 24–26, 59–60,
  71, 119; practices 2, 5–6, 8–9,
  11–15, *23*–24, 27–31, 33, 35–45, 48,
  54, 57–58, 60–61, 64–66, 68–73, 89,
  92, 102, 135, 137; technology 1–5,
  17–18, 20, 22–*23*, 26–30, 35–37,
  39–40, **53**–55, 60–62, 71, 75, **77**,
  80–81, 84, 89–*94*, 99, **101**, **117**, *120*,
  126, **130**, **136**
Epistemic 5–6, 8–9, 11–15, 20, *23*, 26,
  32, 36, 38, 42–45, 49, 74, 89–90,
  92, 100, 119, 135; goals 4–6, 14–15,
  29, 31, 38, 47, 51, 56, 59, 65, *80*,
  88; practice 5–6, 8–9, 11, 15, 18,
  24–26, 29, 38, 40, 43, 52, 54–56,
  58, 65, 68, 73, 75, 79, 81, 89, **91**,
  96, 99, **101**, 116–119, 129–**130**,
  136; practices 2, 5–6, 8–9, 11–15,
  23–24, 27–31, 33, 35–45, 48, 54,
  57–58, 60–61, 64–66, 68–73, 89,
  92, 102, 135, 137
Evaluate 14, 33–34, 52, 54, 64–65,
  69–70, 74, 79, 84, 90, 92, *120*, 127,
  133, 138; evaluated 49, 66, 74, 133,
  138; evaluating 9, 12, 29–30, 34,
  64–66, 74–75, 100, 119, 129, 135;
  evaluation 15, 30, 32, 41, **53**, 64, 66,
  70, 74–75, **77**, 90–**91**, **101**, 116–**117**,
  130, 133, **136**, 138
Evidence 5, 8–9, 12–14, 21, 29–32,
  44–**45**, 49–51, 73, 76, 81, 90, 100,
  119–**130**, 134, 138; evidence-based
  reasoning 9; evidence-informed 138
Experience 4, 7–8, 11, 14–15, 20–22,
  25–26, 28, 31, 35, 45–46, 49, 52, 74,
  81, 89–90, 96, 98, 102, 110, 116,
  126, 129, 137–139

Experiences 4, 7, 13–14, 16–17, 21–22,
  28–29, 31–32, 37–38, 40, 42, **44**, 46,
  48, 52–54, 57–58, 71, 73, 76–**77**, 79,
  88–**91**, 99, **101**–102, 109, 115–118,
  128–131, **136**–138; everyday 12–13,
  21–22, 28, 33, 48, 58, 133; learning
  2, 4, 6–8, 10–11, 13–18, 21, *23*–36,
  38, 40–43, 45–46, 48, 51–52, 54–60,
  63–67, 70–74, 79, 81, 89–90, 92–*103*,
  107–109, 115–116, 118–119,
  122–123, 128, 131, 134, 137–139;
  lived 31, 46
Experiment 37, **39**, 51; experiments
  38–**39**, 49, 51, **53**, 57, 74, **77**, **91**,
  101, *103*, 111, **117**, **130**, **136**
Expert 2, 46, 49, 79, 81, 131–**136**, 139;
  expertise 2, 10, 37, 70, 79, 90, 102;
  experts 1, 49, **53**, **77**, **91**, **101**, **117**,
  130–131, 133–**136**; knowledge 2,
  4–24, 28, 33–34, 36, 38, 40, 42,
  44–50, 52–54, 57–65, 68, 73–**77**, 79,
  81, 89–**91**, 100–102, 108, 112,
  116–118, *120*–122, 125–126,
  128–**130**, 133–137
Explain 5, 10, 13–14, 22, 50, *80*, 97,
  119, 139; explained 14, 88, 111, 134;
  explaining 5, 24, 28–29, **39**, 52, 84;
  explanations 8–10, 12, 28–30,
  44–**45**, 52, 73, 138
Extended 2, 9–10, 28, 31–32, 36, **53**,
  73, **77**, 90–**91**, **101**, **117**, **130**, **136**

Factors 4, 20, *23*, 25, 32–33, 37, 40, 49,
  63, *80*, 116, 119, 121
Familiar 2, 12–13, 20–21, 38, 48–51,
  73, 75, 83, 89, 99, 102, 119,
  128–129; standard 97
Faraway village 106
Feasibility 38–**39**, 69, 126, 134
Feedback 32, **39**, 51, **53**, 58, 63–66,
  69–70, 74–**77**, 90–**91**, 100–**101**,
  116–118, 124–126, 129–131,
  133–134, **136**; peer 22, 27, 43, 51,
  63, 65–70, 74, 99, 116, 133, 139
Findings 14, 111, 134
Fixed number 100
Fixed quantity 95
Fixed time 95
Floating devices 99
Fluid dynamics *103*, 110
Force 86, 96, *132*
Formal reasoning 11
Formative 67–70

Index 145

Foundational Knowledge 36
Frame 17, 72
Framework 14–15, 18–20, 24–26, 28, 31–32, 35–38, 40–**44**, 54–55, 67, 71–72, 74–**77**, *80*, 119–*120*, *132*, 137–139; user-centric 31, 35–36, **39**, **53**, 75–**77**, **91**, **101**–*103*, 116–**117**, 119, 129–131, 135–**136**, 139
Free flow buffet 95
Function 49, 99–100

Generalising 9–10, 28, **44**
Geometry 79, 126; geometer 119–*120*, 126–128
Goals 4–6, 14–15, 29, 31, 38, 47, 51, 56, 59, 65, *80*, 88
Graph 51, 61, *103*–104, 106–109, 115–116; graphs 106; theory 8, 13–14, 26, 55, 61, 71, *103*–104, 106–109, 115–116; weighted 106
Graphic 105, 116
Greedy algorithms 106
Groups 2, 31–32, 37, 51, 65, 67, 69, 81–82, 84–85, 87, 90, 96–97, 108, 111, 121–122, 124, 133–134; breakout 98
Guidance 47, 126
Guided-discovery 11

Habits of mind 9, 25
Hands-on 49, 81, 116
Health 19–20, 37, 47, 76, 119–121, 129, 131–134
Health-related 134
Hearts 102, 104
Houses 109, 112
Humanistic 35–36, 102–118; knowledge 2, 4–24, 28, 33–34, 36, 38, 40, 42, **44**–50, 52–54, 57–65, 68, 73–**77**, 79, 81, 89–**91**, 100–102, 108, 112, 116–118, *120*–122, 125–126, 128–**130**, 133–137; perspective 12, 14, 22, 24, 29, 65, 70, 75, 99, 102
Hypothesis-testing 30

Identifying 5, 17, 20, 26, 32–33, 35, 38–**39**, 41, 71, 119, 133, 138
Impact 32, 47, **53**, 70, **77**, **91**, **101**, **117**, **130**, *132*, **136**
In-depth 89, 112, 116, 134
Inclusion *23*, 125, 139
Inductive reasoning 10

Industrial **53**, **77**, **91**–92, **101**, **117**, **130**, **136**
Inquiry 1–29, 31–33, 38, 41–78, 82–84, 87–92, 96, 98–119, 121–131, 133–139; field 49, 73, *120*, 129; general 8, 12–13, 55, 62, 95; inquiring 5, 61–62; inquiry-based learning 8, 25; inquiry-based mathematics task 106; mathematical 5–6, 9–14, 18, 20, 24–27, 29, 37–38, 42, **44**–**45**, 47, 76, 79–*80*, *103*, 109, 116, *120*, 126; mathematics 2–5, 7, 9–12, 14, 17–18, 20, 22–*23*, 25–30, 35, 37, **39**–40, 42, **44**, 49–51, **53**, 55, 71, 73–74, 76–**77**, 79–81, 90–*94*, **101**, *103*, 106, 108–109, 115, **117**, *120*, 126, **130**, *132*, **136**–137; science 2–10, 12–15, 17–18, 20, 22–31, 35, 37–43, 48–51, **53**, 55–56, 60, 71, 73–74, 76–81, 89–*94*, **101**, *103*, 115, **117**, **130**, *132*, 134, **136**–137; scientific 8–9, 11–15, 24–25, 27–30, 32, 37–**39**, 41, **44**–47, 55, 89, *103*; solution-centric 31–34, 36–41, **53**, 74–**77**, 90–92, 99–**101**, **117**, 119, **130**, 133, 135–**136**, 139; user-centric 31, 35–36, **39**, **53**, 75–**77**, **91**, **101**–*103*, 116–**117**, 119, 129–131, 135–**136**, 139
Instruction 17, 55, 57–59, 81, 102, 138; direct 37, 51, 81, 89, 96, 116; instructional framework 15, 19, 28, 31–32, 37, 40–42, 54, 75–**77**, *80*, 119–*120*, *132*, 137–139; instructional strategies 81; whole-class 105–106, 116
Integrated knowledge 59
Integrated STEM 2, 4, 6, 9, 11–15, 17–18, 21–79, 88–89, 99–100, 102–*132*, 134–135, 137–139; classroom practices 38, 72; classrooms 8, 24–25, 27, 32, 43, 54, 72–78, 82, 95, 99, 104, 115, 131, 137; disciplinary 4–6, 8–17, 22–24, 27–30, 33, 36–37, 42–46, 49–50, 55, 57–59, 61, 63, 65, 68–69, 71, 74, 79, 90, 102, 135; education 2–4, 6–8, 14, 18, 24–29, 41, 52, 55–56, 59, 64, 70–71, 78, 128; frameworks 17, 24, 27–28, 41, 71, 78; inquiry 1–29, 31–33, 38, 41–78, 82–84, 87–92, 96, 98–119, 121–131, 133–139;

Index

integrate STEM 17, 58, 65; learning 2, 4, 6–8, 10–11, 13–18, 21, *23*–36, 38, 40–43, 45–46, 48, 51–52, 54–60, 63–67, 70–74, 79, 81, 89–90, 92–*103*, *107*–109, 115–116, 118–119, 122–123, 128, 131, 134, 137–139; learning experiences 16, 28–29, 32, 38, 40, 42, 48, 52, 73, 79, 90, 102, 109, 116, 118, 128, 131, 137–138; lesson 4, 27, 29, 31–32, 34, 37–38, 40, 43–45, 51, 56, 58–59, 69, 72, 76, 79–*80*, 82, 84–85, 89–90, 92, 95, 100, 102, 106, 119–*120*, 128, 131–133, 137; problem-solving 4, 9, 14, 16–17, 30–33, 36–37, 45–46, 48–49, 52, 55, 58, 60, 66, 68–69, 74–75, 79, 90, 100, 102

Inventive 75, 138; inventions 138

Iterative 8, 29–30, 34, 45–46, 58, 63, 67–69, 74–75, 90, 100

Justification 12, 32–33, 43–**44**, 73, 119, 129; justification(s) 43–**44**, 73, 129; justified 134; justifying 9, 28, **44**–**45**, 135

Knowledge 2, 4–24, 28, 33–34, 36, 38, 40, 42, 44, **44**–50, 52–54, 57–65, 68, 73–**77**, 79, 81, 89–**91**, 100–102, 108, 112, 116–118, *120*–122, 125–126, 128–**130**, 133–137; map 14, 26, 71, *103*–104, 109, 116, 121–122, 131; referent-centred knowledge 20–21; subject matter 7, 20; subject-matter 14, 21–22, 25, 79, 102, 116, 122

Language 7, 73
Laws of nature 5
Learner 4, 64, 70, 139; profile 139
Learners 7, 9–10, 17, 27, 43, 46, 48–49, 52, 54, 73–75, 92, 138–139; Profiles 4, 54
Learning 2, 4, 6–8, 10–11, 13–18, 21, *23*–36, 38, 40–43, 45–46, 48, 51–52, 54–60, 63–67, 70–74, 79, 81, 89–90, 92–*103*, *107*–109, 115–116, 118–119, 122–123, 128, 131, 134, 137–139; active 7–8, 11, 21; contexts 4, 9, 19, 24–26, 32, 35, 40–41, 48, 50, **53**, 55, 65, 70, 73, 75, **77**, **91**, **101**–102, **117**, **130**, **136**–137; goals 4–6, 14–15, 29, 31, 38, 47, 51, 56, 59, 65, *80*, 88; intention(s) 31–32, 59; intentions 52–**53**, 56–57, 72, 75, **77**, 79, **91**, 97, **101**, **117**, 119, **130**, **136**; needs 2, 5, 7, 14, 19, 31, 35–36, *39*, **44**, 46, **53**, 57–60, 75, **77**, **91**, **101**, 104, 116–**117**, **130**, **136**; opportunities 8, 14–15, 22, 43, 46, 48–50, 52–**53**, 58, 72–75, **77**, 79, 81, 90–**91**, 99–102, 109, 116–**117**, 121–123, 125, 128–**130**, 133–134, **136**–138; outcome(s) 36, 38; outcomes 12, 20, *23*, *39*, 54–55, 70, 79, 92, *103*, 119, 134, 138–139; profiles 4, 54

Lesson 4, 27, 29, 31–32, 34, 37–38, 40, 43–45, 51, 56, 58–59, 69, 72, 76, 79–*80*, 82, 84–85, 89–90, 92, 95, 100, 102, 106, 119–*120*, 128, 131–133, 137; intention 16, 31–32, 49, 58–59, 76, 102, 135; plan 40, 59, 73, 79–81, 92, 102, 131; problem-centric 31–33, 36–37, *39*, 41, **53**, 73–**77**, 79, 89–**91**, **101**, **117**, 119, 121, 129–**130**, 133, 135–**136**, 139

Lessons 15, 29–31, 33, 36–38, 40–41, 43–48–50, 59, 70, 72–74, 76, 79, 92, 99, 116, 139

Light detection 89
Light scattering 89
Limitations 20, 24, 31, 51, **53**, **77**, **91**–92, **101**, **117**, **130**, **136**
Linear 20
Literacy 18, 24–26, 59–60, 71, 119
Litter *93*, 95, 97–100; picker 98, 100
Living environment 116
Logic 12, **44**; logical deduction 42; logical reasoning 10, 12, 14, 28
Lunch hour 95

Making 1–27, 31, 43–**44**, 50–**53**, 55, 72–75, **77**, *80*–81, 83–84, 89, **91**, *93*–94, 96, 98–102, 110, 115, **117**, 126–128, **130**–131, 135–**136**, 138
Making 1–27, 31, 43–**44**, 50–**53**, 55, 72–75, **77**, *80*–81, 83–84, 89, **91**, *93*–94, 96, 98–102, 110, 115, **117**, 126–128, **130**–131, 135–**136**, 138
Man-made world **44**
Map 14, 26, 71, *103*–104, 109, 116, 121–122, 131; mapped **39**–40, 54, 119; mapping 37–38, 40, 43, **53**, 76, 102–*103*, 135, 138; reading 104, 116
Materials 50–**53**, 58, 61, **77**, 79–82, 84, **91**, *93*–95, 99, **101**, **117**, *120*, 123,

Index 147

125–126, **130**, 133, **136**, 138; budget *80*, 82, 85, 90, 95, 100; material 21, 30, 43, 46, 50, 72, 89, 95–96, 99, 116, 126, 137
Math 24, 27; mathematical 5–6, 9–14, 18, 20, 24–27, 29, 37–38, 42, **44–45**, 47, 76, 79–*80*, *103*, 109, 116, *120*, 126; mathematical communication 109; mathematical modelling 20, 76, 109, 116; mathematical thinking 9–10, 18, 25; mathematicians 9–10, **45**, **53**, **77**, **91**, **101**, **117**, **130**, **136**; mathematics 2–5, 7, 9–12, 14, 17–18, 20, 22–*23*, 25–30, 35, 37, **39**–40, 42, **44**, 49–51, **53**, 55, 71, 73–74, 76–**77**, 79–81, 90–*94*, **101**, *103*, 106, 108–109, 115, **117**, *120*, 126, **130**, *132*, **136**–137; mathematics and computational thinking 29; mathematics-engineering 81
Meaningful 4, 9, 13, 15–17, 21–22, 25, 31–32, 46, 48–49, 54, 69, 73–74, 79, 100, 104, 139
Measured 37, 57, 109, 134
Measurements *23*, **44**, *80*, 119, *132*
Mechanism 5, 96, 99
Mentimeter 121
Meta knowledge 36
Metacognitive 46, 64
Methods 10, 21–22, 57–59, 63–68, 70, 119, 133, 138
Microphones 95
Mimicking seeds 92
Model 4, 6, 13–17, 21, *23*–25, 32, 35, 37–38, 40, 51–52, 55, 57–58, 63, 69, 76, *80*–81, 89, *103*, 109, 112–116, 118–119, 126–128, 133–134; modelling 20, 43, 50, 52–**53**, 58, 72, 76–**77**, 83–86, **91**, 100–**101**, *103*, 109, 115–**117**, 127, **130**, 134, **136**; models 2–4, 6, 9, 12, 14–18, 25–26, 29–31, 36, **39**, 51–**53**, 55, 57–58, 62, 70, **77**, **91**, **101**, **117**, 126, **130**, 134, **136**, 138; physical *23*, 37, 50, 81, 89, 116, *120*, 131, 133–134; scaled 63, 112–115
Modern 1, 7, 131
Modify 124
Momentum **45**, 75, 90, 100
Monitor *23*, 68
Mountain 20, 61, 102, 104, 106, 112, 114, 116, 118; mountainous 19–20, 102–104, 113, 115–116;

mountainous regions 20, 102, 104, 115–116; mountains 61, *103*, 105, 112, 114, 116; pipelines 102, 110–111; remote 13, 19–20, *103*–104, 115–116
Mr Jet 95, 97–98, 100
Mr Pu 95, 97–98, 100
Mr Teerute 104, 110–112
Multi-dimensional 2, 10, 46
Multi-faceted 57, 59
Multi-level 57
Muscle 96

National standards 28
Natural 5, 7, 28–30, **44**, 46, 121
Natural phenomena 28–29, 46
Needs 2, 5, 7, 14, 19, 31, 35–36, **39**, **44**, 46, **53**, 57–60, 75, **77**, **91**, **101**, 104, 116–**117**, **130**, **136**
Negotiations 100
Next Generation Science Standards (NGSS) 28; Next Generation Science Standards 9, 26, 28, 41; NGSS 9, 26, 28–30, 41
Numbers 5

Objective 30, 63–65, 67, 74
Objectives 36, 57, 59, 76
Objects 10, 14, **53**, 73, 75, **77**, **91**, 99, **101**, **117**, **130**, **136**, 138; common 2, 4, 9, 12–13, 16, 19, 21, 32, 42, 52, 57, 65, 73, 85, 90, 96, 99, 137–138
Observation 10, 43, 55, 71, 73, 129, 137–139; making 1–27, 31, 43–**44**, 50–**53**, 55, 72–75, **77**, *80*–81, 83–84, 89, **91**, *93*–*94*, 96, 98–102, 110, 115, **117**, 126–128, **130**–131, 135–**136**, 138; observations 10, 32, **44**, 51, 61–62, 66, 73, 81, 90, 100, 119, 137–139; observing 10, 90
Online research 81
Open-ended questions 121
Opportunities 8, 14–15, 22, 43, 46, 48–50, 52–**53**, 58, 72–75, **77**, 79, 81, 90–**91**, 99–102, 109, 116–**117**, 121–123, 125, 128–**130**, 133–134, **136**–138
Opportunity 70, 98, 100, 128; prototyping **39**, 114–115, 123; refinement 51, *94*, 99, 125, 129, 134
Outcomes 12, 20, *23*, **39**, 54–55, 70, 79, 92, *103*, 119, 134, 138–139
Overlaid 95

148 Index

Paper-and-pencil tests 57
Peer 22, 27, 43, 51, 63, 65–70, 74, 99, 116, 133, 139; peer critique 22, 43, 68, 74, 99, 116, 133; peer-reported 65–70; peers 24, 30, 52, 54, 58, 63–65, 69, 73, 81, 83, 88, 95, 100, 122, 124–125, 129, 134, 138
perception 4, 8, 64, 122
Performance 25, 32, 57, 64, 100; tasks 9–11, 25, 27, 57, 95, 106–109, 134
Perspectives 4, 10–11, 26, 29, 36, 66, 74, 104
Persuade 30, **44**, 54, 129, 134, 138
Philosophy 4, 7, 57–58, 60
Pincer 96, 100
Pipeline system 110, 112, 114–115; pipelines 102, 110–111; pipes 50, 61, *103*, 106, 111, 116, 125; piping layout 105
Policies **53**, 64, **77**, **91**, **101**, **117**, **130**, **136**
Potential 24, 34, 36–37, 64, 66
Practices 2, 5–6, 8–9, 11–15, *23*–24, 27–31, 33, 35–45, 48, 54, 57–58, 60–61, 64–66, 68–73, 89, 92, 102, 135, 137; pedagogical 7, 11–12, 15, 21–22, 26, 29–30, 43–45, 72, 89, 99, 116, 118–119, 129; practice knowledge 73; professional 26, 42, 54, 73; science and engineering *3*, 5, 9, 28–31, 41, 50, **53**, **101**, **117**; students' 2, 7–8, 13–15, 19–21, *23*, 27, 31, 34, 36–38, 40–41, 46–52, 56–59, 63–66, 68–70, 72–74, 81–82, 84, 88–90, 95, 99–100, 102, 104, 115–116, 119, 121–122, 125, 128–129, 131, 137–139
Prediction 30, *94*, 99
Preferences 35, 66, 75
Present 14, 38, 40, 43, 46, 49–50, **53**–54, 65, 72–73, 76–**77**, **91**–92, 95, **101**, **117**, 119, **130**, 134, **136**–138; presentation 43, **45**, 49, **53**, 68, 72, **77**, 90–**91**, 95, 98, **101**, **117**, 121–125, 128, **130**, 134, **136**; presentations 65, 81, 95, 129, 134
Pressure 49–50, 61, *94*, *103*, 105, 110–111, 116, *132*; fluid *103*, 110, 116
Problem 1, *3*–7, 9–11, 13–17, 19–50, 52–55, 57–58, 60–63, 66–69, 71–**77**, 79–86, 88–**91**, *93*–95, 100–*103*, 105–106, 108–**117**, 119–121, 127–**130**, *132*–139; definition 26, 47, 63, 67–68; identification 33, 129; practical 8, 12–14, 21–22, 28, 32, 37–38, 40, 42, 47, 58–59, 62, 68, 115, 118, 129; problem solving 1, *3*, 5–6, 11, 14, 17, 20–22, 25–28, 31–32, 37, 40–41, 48–50, 52, 54, 57–58, 60, 62, 68–69, 71, 74, 79, *93*–95, *120*, 129, 137; problematising 73, 129; problem-centred knowledge 20–21, 24, 54; problem-centric variant 32, 36, 76, 89, 121; problem-centric 31–33, 36–37, **39**, 41, **53**, 73–**77**, 79, 89–**91**, **101**, **117**, 119, 121, 129–**130**, 133, 135–**136**, 139; problem-solving cycle **45**, 75, 90, 100; problem-solving skills 58, 68–69, 74; problem-solving 4, 9, 14, 16–17, 30–33, 36–37, 45–46, 48–49, 52, 55, 58, 60, 66, 68–69, 74–75, 79, 90, 100, 102; solve 1, 9–11, 14–15, 17, 20–21, 32, 36, 42, **44**–45, 47, 49, 52, 61, 63, 68, 74, 79, 81, 86, *94*, 106, 108, 127; solvers 43, 52–**53**, 72, 74, **77**, 83–86, 90–**91**, 100–**101**, 109, **117**, 127, **130**, 134, **136**, 138; solving 1, *3*–6, 9, 11, 14–17, 20–22, 25–33, 36–37, 40–41, **44**–46, 48–50, 52, 54–55, 57–58, 60, 62, 66, 68–69, 71, 74–75, 79–80, 90, *93*–95, 100, 102–*103*, *120*, 129, *132*, 137
Problem-based 14; group 2, 26, 32, 43–45, 49, 51–52, 58–59, 63–67, 70, 74–75, *80*–81, 85–87, 89–90, *93*–95, 98, *103*, 105, *120*–124, 127–129, *132*–134, 138
Procedural 38, 40, 43–45, 49, 62, 75, 90, 100, 116
Process *3*, 5, 7–8, 10, 12–14, 17, 21–*23*, 30–31, 33–36, 40–41, 45–46, 49–52, 55, 58, 62–63, 67–68, 70, 73–75, 79–81, 85, 87, 90, *93*–94, 97, 99, 109, 114, *120*, 123, 126, 131, 134
Processes 4–5, 7, 9–10, 12, 14, 17, 20, 27, 29–30, 32, 34–36, 47–52, 58, 63, 68–69, 74, 82, 89, *94*, 128–129, 135; flaws 97
Product 5, 32, 37, 40, **53**, 58, 61–62, 68, **77**, **91**, **101**, **117**, **130**, **136**; features 4, 6, 16–17, 19, 28, 31–32, 42–43, **53**, 61, **77**, **91**, **101**, **117**, 129–**130**, 133, **136**; human-made 5; product

Index 149

5, 32, 37, 40, **53**, 58, 61–62, 68, **77**, **91**, **101**, 117, **130**, **136**; product-focused 58; productive 8–10, 43, 52, 55, 62, 71, 74, 79
profile 139
Progress 30, 48, 51, 65, 90
Projects 16, 65, 95, 97; student 7, 13, 21, 24–25, 31, 38, 45, 52, 54, 58–67, 70–71, 81, 97, 100, 129, 138
Proposals 119
Pros 134
Protocols 59
Prototype 22, 51–52, 63, *80*–81, 86–87, 90, 96–98, 100, 115, *120*, 123, 126–128, 138; crab claw *93*, 97; prototypes 22, 30, **39**, 51–**53**, 63, 69–70, 75, **77**, 81, 88–89, **91**, 98, **101**, **117**, 119, 125–127, 129–**130**, **136**–138

Quantities 5
Question 4, 6, 9, 13, 28–29, 82
Questioning 8, 41, 84; Teaching and Learning 6, 8, 15, 27, 46, 55, 57, 73, 116

Raft *94*, 98–99; foldable *94*, 98–99
Readiness 32, 36, 38, 40, 48
Real-world *3*, 14–16, 18–19, 27–28, 31, 33, 42–**44**, 46–49, **53**, 57, 72–73, **77**, 82, 88–**91**, 99, **101**–102, 106, 115, **117**, 121, 128–**130**, 133–137
Realistic 19, 34, 48, 54
Redefining 55, 68
Refining 33–34, **39**, 58, 70, 75
Relationships 5, 16–17, 20, 33, 38, 40, 58, 61–62, 111
Representative 52
Research 14, 24–27, 41, 49, 54–56, 63, 67–71, 78, 81–82, 95, 119, 133–134, 139; researchers 16, 73, 129, 139; researching 38–**39**, 68–70
Resource acquisition 81
Resource limitation 95
Responsibilities 121
Result 32–33, 40, 46–47, 49, 102, 138
Reviewed 123
Revise 63, 69
Roles 12, 16–17, 43–**44**, 49, 73, 121
Room 95
Round-robin 70
Rubric 62, 65–67; rubrics 59, 65
Rules 45–46, **53**, **77**, **91**, **101**, **117**, **130**, **136**

S-T-E-M Quartet instructional framework 19
Samples 98; real-life 4, 21, 35, 46, 76, 98
Sawtooth 96
Scaffold 19, 50, 72
Scale *23*, 85, 116, *120*, 126
Scenario 97
School 2, 6, 9, 13, 19–21, 24–27, 37, 41, 46, 48, 62, 71, 79–82, 86–87, 89, 95, 99, 102, 119–121, 123, 127–129, 131, 138; hall 95–96; subject 6–7, 14–15, 20–22, 25, 79, 102, 116, 122, 137
Science 2–10, 12–15, 17–18, 20, 22–31, 35, 37–43, 48–51, **53**, 55–56, 60, 71, 73–74, 76–81, 89–*94*, **101**, *103*, 115, **117**, **130**, *132*, 134, **136**–137; physical *23*, 37, 50, 81, 89, 116, *120*, 131, 133–134; science-focused 29–30; scientific 8–9, 11–15, 24–25, 27–30, 32, 37–**39**, 41, **44**–47, 55, 89, *103*; scientific knowledge 8–9, 14, 28, 89; scientific method 29; scientists 8, **44**–45, **53**, **77**, **91**, **101**, **117**, 129–**130**, **136**; traditional 28, 57, *93*–*94*
Score 66, 69, 99
Self 21, 52, 54, 63–65, 67; report 41, 63; reported 1, 12, 21, 29, 57, 63–70
Sensitivities 102
Sharing 40, 62, 68, 85, 106, 121
Similarities 4, 12–13, 15, 24, 29, 42–43, 48
Sketch 63, *120*, 123
sketching 95–96
Skills 2, 9, 13, 15–17, 20, 22, 26–29, 31, 33–34, **45**, 47, 49, 52, 54–55, 57–58, 60–62, 64–65, 68–69, 73–75, 79–82, 89–90, *93*–*94*, 96, 102, 121, 137
Snippets 104, 115
Social 4, 7, 9, 14, 19, 25, 33, 42–46, 52–**53**, 70, 72–73, 75, **77**, 90–**91**, 100–**101**, 116–**117**, 129–**130**, **136**, 138; interactions 2, 47, 52–**53**, 75–**77**, 82, **91**, 95, **101**, 104, **117**, 121, **130**, 133, **136**, 138
Software **39**, 82, *120*, 126
Solution 7, 17, 20, 22, 31–**44**, 46–47, 50, **53**, 58, 64, 66, 73–**77**, 90–92, 99–**101**, 105–106, 109, 111–112, 114, **117**, 119–*120*, 127, 129–**130**, *132*–**136**, 139; integrated 2–4, 6, 9,

11–15, 17–18, 21–79, 88–89, 99–100, 102–135, 137–139; piping 105; solution-centric 31–34, 36–41, **53**, 74–**77**, 90–92, 99–**101**, **117**, 119, **130**, 133, 135–**136**, 139; solutions 2, 5, 15, 18, 21–*23*, 28–34, 38–**39**, 46–47, 49–**53**, 58, 63, 66, 68, 70, 73–75, **77**, 79, 84, 86, **91**–92, **101**–102, 115, **117**, 121–122, 125, 127, 129–**130**, 133–134, **136**, 138–139; testing 7, 14, 29–30, 33, **39**, 42–43, 51, 63, 68–69, 74–75, 81, 84, 87, 90, 97–100, 116, 134
Sound system 95; integrated 2–4, 6, 9, 11–15, 17–18, 21–79, 88–89, 99–100, 102–135, 137–139
Specialising 9–10, 28
STEM 1–*80*, 82, 88–92, 99–139; activities 7, 14, 27–28, 32, 38–**39**, 41, 47–48, 57, 59, 61, 63, 65–67, 71, 74, 79, 95, 102, 119, 138; activity 21, *23*, 25, 37–38, 48–51, 58–59, 61–62, 65, 70, 76, 81, 89–90, 99–100, 102, 109, 119, 121, 128–129, 131, 133–135, 138; approach 6–7, 11, 14, 17, 26, 37, 48, 92, 99, 139; curricula 2, 4, 10, 21, 25, 28, 58; disciplines 2–9, 12–17, 19–20, 22–24, 28–29, 33–35, 37–38, 42–43, 48–50, 58, 74, 79, 102, 118, 137; education 2–4, 6–8, 14, 18, 24–29, 41, 52, 55–56, 59, 64, 70–71, 78, 128; inquiry 1–29, 31–33, 38, 41–78, 82–84, 87–92, 96, 98–119, 121–131, 133–139; learning 2, 4, 6–8, 10–11, 13–18, 21, *23*–36, 38, 40–43, 45–46, 48, 51–52, 54–60, 63–67, 70–74, 79, 81, 89–90, 92–*103*, *107*–109, 115–116, 118–119, 122–123, 128, 131, 134, 137–139; lesson 4, 27, 29, 31–32, 34, 37–38, 40, 43–45, 51, 56, 58–59, 69, 72, 76, 79–*80*, 82, 84–85, 89–90, 92, 95, 100, 102, 106, 119–*120*, 128, 131–133, 137; lesson phase 58–59; lessons 15, 29–31, 33, 36–38, 40–41, 43, 48–50, 59, 70, 72–74, 76, 79, 92, 99, 116, 139; practices 2, 5–6, 8–9, 11–15, *23*–24, 27–31, 33, 35–45, 48, 54, 57–58, 60–61, 64–66, 68–73, 89, 92, 102, 135, 137; problem 1, *3*–7, 9–11, 13–17, 19–50, 52–55, 57–58, 60–63, 66–69, 71–**77**, 79–86, 88–**91**, *93*–95, 100–*103*, 105–106, 108–**117**, 119–121, 127–**130**, *132*–139; problem solvers 43, 52, 72, 74, 83–86, 90, 100, 109, 138; programmes 14, 57–59, 71; quartet 15, 19, 27–28, 31–32, 35–38, 40–42, 54, 71–72, 75–78, *80*, 92, 119–*120*, *132*, 137–139; quartet framework 35–36, 38, 54, 72; quartet instructional framework 15, 19, 28, 31–32, 37, 40–42, 54, 75–**77**, *80*, 119–*120*, *132*, 137–139; unit 41, *80*, 104, 108
Stimuli 105
Strategies 38, 46, 48–50, 67, 70–71, 79, 81, 129
Structural 126
Structuredness 61
Student 7, 13, 21, 24–25, 31, 38, **45**, 52, 54, 58–67, 70–71, 81, 97, 100, 129, 138
Students 2, 7–8, 10–*23*, 25–27, 30–38, 40–41, 43–59, 61, 63–66, 68–70, 72–**77**, 79, 81–92, 95–102, 104–106, 108–119, 121–131, 133–139; abilities 32, 37–38, 47, 49, 60, 64, 70, 74, 134; directed 81, 89, 138; learning 2, 4, 6–8, 10–11, 13–18, 21, *23*–36, 38, 40–43, 45–46, 48, 51–52, 54–60, 63–67, 70–74, 79, 81, 89–90, 92–*103*, *107*–109, 115–116, 118–119, 122–123, 128, 131, 134, 137–139
Subject 6–7, 14–15, 20–22, 25, 79, 102, 116, 122, 137
Sustainable 46–47, 56, 66, 88, *132*
Symposium 137

Teaching 4, 6–8, 11, 15, 24–27, 37, 41, 46, 52, 55, 57, 71, 73, 76, 78–79, 81–82, 116, 121; strategies 38, 46, 48–50, 67, 70–71, 79, 81, 129; teacher 6, 19, 22, 24, 26–27, 37–38, 40–41, 49–50, 55–56, 63–69, 71, 79, 81–84, 89, 96, 131; teacher aide 96; teacher-facilitated 81; teacher-subjective 63–67, 69; teachers 2–4, 6–7, 11, 13, 15, 21, 24–27, 30–31, 36–38, 40–45, 48–52, 54–55, 59, 62, 64, 66–74, 76, 79, 81–84, 88–90, 92, 95, 99–100, 102, 104, 116, 121–123, 125–126, 128, 133–135, 137–139; teams 2, 46, 52,

Index   151

79, 90, 95, 98, 138; vignette 68–70, 76, 78–79, 82, 88–89, **91**–92, 95, 99–102, 104, 115–119, 121, 128–131, 133–**136**
Teaching and Learning 6, 8, 15, 27, 46, 55, 57, 73, 116; science 2–10, 12–15, 17–18, 20, 22–31, 35, 37–43, 48–51, **53**, 55–56, 60, 71, 73–74, 76–81, 89–*94*, **101**, *103*, 115, **117**, **130**, *132*, 134, **136**–137
Team 65, 73, 79–**91**, 97, 100, 102, 104, 121, 129, 133, 137; teamwork 64
Technology 1–5, 17–18, 20, 22–*23*, 26–30, 35–37, **39**–40, **53**–55, 60–62, 71, 75, **77**, *80*–81, 84, 89–*94*, 99, **101**, **117**, *120*, 126, **130**, **136**; science 2–10, 12–15, 17–18, 20, 22–31, 35, 37–43, 48–51, **53**, 55–56, 60, 71, 73–74, 76–81, 89–*94*, **101**, *103*, 115, **117**, **130**, *132*, 134, **136**–137; technologies *23*, **53**, 55, 75, **77**, **91**, **101**, **117**, **130**, **136**
Testing 7, 14, 29–30, 33, **39**, 42–43, 51, 63, 68–69, 74–75, 81, 84, 87, 90, 97–100, 116, 134; ideas 2, 4, 6–8, 10–12, 14–15, 22, 29–30, 32, 36–37, 42–50, 52–55, 58–59, 61–63, 65–66, 68–70, 72–74, 76–**77**, *80*, 82, 88–89, **91**, *93*–95, 97–**101**, *103*, 105–106, 108–110, 115–**117**, *120*–125, 127–134, **136**–139
Time 8, 15, *23*, 32–33, 38, 50–51, 69, 73, 79–81, 87, 95, 98–99, 108, 110, 113, 115–116, 121, 124, 137–139; allocation 81
Tool 10, **39**, 68, 81, 89, 95, 138
Transient 47
Trends 38–40, 61–62

Understanding 4–7, 10, 12–16, 20–22, 25–26, 28, 31–33, 35–36, 42, **44**–46, 49–50, **53**, 57, 60, 64, 73, 75–**77**, 84, 89, **91**, 99, **101**, 116–**117**, 129–**130**, 133–134, **136**, 138
United Nations Sustainable Development Goals (UNSDG) 88
Unoccupied tenor 95
Usefulness 134
User 31, 35–37, **39**, 42, **44**, **53**, 75–**77**, **91**, **101**–104, 116–**117**, 119, 129–131, 135–**136**, 139; expectations **53**, **77**, **91**, 96, **101**, **117**, **130**, **136**; user needs 35–36, **53**, 75, **77**, **91**, **117**, **136**; user-centric 31, 35–36, **39**, **53**, 75–**77**, **91**, **101**–*103*, 116–**117**, 119, 129–131, 135–**136**, 139; STEM 1–*80*, 82, 88–92, 99–139; variant 1, 32–36, 41, 54, 72–73, 76, 89–90, 92, 121, 139

Values 4–5, 12, 29, 59, 102–118
Variants 28, 32, 36–38, 40, 54, 72, 75–**77**, 90, 100, 118–119, 133, 138
Videos 104–105, 115–116
Vignette 68–70, 76, 78–79, 82, 88–89, **91**–92, 95, 99–102, 104, 115–119, 121, 128–131, 133–**136**
Villagers 102–104, 115–116
Visual representations 4, 126

Water 19–20, *23*, 37–38, 48–51, 56, 61–62, 69, 79–89, *94*, 98–99, 102–106, 108, 110–112, 116, 125, 134
Workability **53**, **77**, *80*, 89, **91**, **101**, **117**, **130**, **136**, 138

Taylor & Francis eBooks

www.taylorfrancis.com

A single destination for eBooks from Taylor & Francis with increased functionality and an improved user experience to meet the needs of our customers.

90,000+ eBooks of award-winning academic content in Humanities, Social Science, Science, Technology, Engineering, and Medical written by a global network of editors and authors.

TAYLOR & FRANCIS EBOOKS OFFERS:

- A streamlined experience for our library customers
- A single point of discovery for all of our eBook content
- Improved search and discovery of content at both book and chapter level

REQUEST A FREE TRIAL
support@taylorfrancis.com

For Product Safety Concerns and Information please contact our EU representative GPSR@taylorandfrancis.com
Taylor & Francis Verlag GmbH, Kaufingerstraße 24, 80331 München, Germany

www.ingramcontent.com/pod-product-compliance
Lightning Source LLC
Chambersburg PA
CBHW051747230426
43670CB00012B/2199